山东省社会科学规划研究项目“文学翻译教学探究
——以海明威早期短篇小说翻译为例”(11CWZJ11)
山东省社会科学规划研究一般项目“两汉外来词汉化研究”(14CWXJ59)
山东省高校人文社科项目“两汉佛典与非佛典外来词对比研究”(J15WD11)

顺　道

Law of Adaptation

顺 之 如 凤　著
韩淑红　吴远庆　译

山东大学出版社

图书在版编目(CIP)数据

顺道/顺之,如凤著;韩淑红,吴远庆译.
—济南:山东大学出版社,2016.11
ISBN 978-7-5607-5644-8

Ⅰ.①顺… Ⅱ.①顺… ②如… ③韩… ④吴…
Ⅲ.①人生哲学—通俗读物 Ⅳ.①B821—49

中国版本图书馆 CIP 数据核字(2016)第 264855 号

责任策划:黄福武
责任编辑:郑琳琳 张申华
封面设计:张 荔

出版发行:山东大学出版社
社 址 山东省济南市山大南路 20 号
邮 编 250100
电 话 市场部(0531)88364466
经 销:山东省新华书店
印 刷:山东华鑫天成印刷有限公司
规 格:720 毫米×1000 毫米 1/16
15 印张 285 千字
版 次:2016 年 11 月第 1 版
印 次:2016 年 11 月第 1 次印刷
定 价:39.00 元

译者前言

《顺道》一书是王培禄（字顺之）和王淑玉（字如凤）合著的一部富含人生哲理的哲学著作，它把哲学引进老百姓的实际生活中。《顺道》一经问世，就引起了社会的广泛关注和众多好评。中国外文局原副总编辑、现《对外传播》总编林良旗先生说："《顺道》是老百姓的哲学，是老百姓贴身的哲学。"《烟台日报》评论说："相信顺道思想必将助推社会、经济、文化等诸多方面的发展，为创建文明城市和建设和谐社会做出贡献。"

《顺道》对大学生的思想政治教育特别具有启发意义。大学生在学习、工作、生活当中，会遇到许多对人生、家庭、社会等感到困惑不解的地方，了解"顺道"理论有利于引导青年学生的实际行动，做到顺应历史、顺应万物、顺应和谐，为和谐社会的建设做出自己的贡献。《顺道》中讲述了许多中国优秀传统文化的精髓，传递的是正能量，它是烟台的文化、民族的文化、世界的文化。

在世界多极化、文化多元化的全球一体化新时代，中国政府也在积极推动文化走出去的战略，这就是当前正在掀起的中国典籍外译的高潮。把中华民族富有特色的文化精髓译成外语，是让世界了解中国、让中国文化参与全球文化交流的一个极好途径，正如典籍英译学科委员会召集人卓振英教授所说："对于弘扬民族文化、促进东西方文化融合、保持中国固有的文化身份来说，典籍英译有着十分重大的现实意义。"季羡林先生也曾说过："把翻译的重点从外译中转变为中译外，不仅需要更多具有开放胸襟和气象的'拿来主义'，而且要有更多大胆的'送去主义'。"鉴于此，把《顺道》翻译成英语、把顺道的理论和思想介绍给世界各国人民，也符合"顺道"的基本思想，是顺应时代潮流、传播中国文化的一次有益尝试，有利于宣传中国文化，和谐世界人民。

从接到原书作者翻译《顺道》的委托，到最终完成全书翻译，已是两年多的时间。翻译过程中得到了徐杉杉、姜丽莉、单庆凤、左通、田利、李明涛、杨运婷、邹鹏、魏来贵、麻文静、崔思燕、崔家慧、翟珊、张萌、初开迪、丁赛男、吕娉、李香君等各位同学以及同事、好友和鲁东大学外国语学院领导的大力支持和帮助，尤其是 2013 年在美国布里奇波特大学访学期间，该校教务长 Stephen F. Healey 博士对部分章节的译文给予了具体的指导。同时，作为鲁东大学“国际化学生事务研究工作室”系列研究成果之一，得到了学校专门经费支持，在此一并表示感谢。虽然最终完成了全书的翻译，但一直都在不断地认真修改、润色、加工。许钧先生说过：“翻译决不仅仅是译者译得好不好的个人的小事，它是关乎到一个民族文化的传承、交流与发展的大事。”因此，当真正把译文交给出版社准备出版时，心情还是惴惴不安。原书作者在书中引用了大量的典故、谚语、古诗词，甚至使用了一些方言，并且提出了自己独特的哲学观点和理论，这些都成为翻译的难点。为解决这些问题，在翻译过程中采取了译者加注的方式，译者查找、参阅了部分著作、期刊和网络资料，恕不能一一列出，同时也不时与原作者进行必要的沟通和交流。尽管在翻译过程中对译文进行了反复推敲和研究，但是由于译者水平有限，不当乃至错误之处在所难免，敬请各位读者批评指正。

韩淑红　吴远庆

2015 年 9 月 17 日

Translators' Foreword

Law of Adaptation, written by Wang Peilu (styled Shunzhi) and Wang Shuyu (styled Rufeng), is a philosophical work that is full of life philosophy. It brings philosophy into ordinary people's real life. *Law of adaptation* has aroused many concerns and praises since it was published. Mr. Lin Liangqi, the former deputy editor-in-chief of CIPG (China International Publishing Group) and the current editor-in-chief of the journal *International Communications said*, "*Law of Adaptation* is a philosophy that belongs to common people. It is a philosophy that is very close to people's life." *Yantai Daily* commented, "The idea of law of adaptation will certainly give an impetus to the development of society, economy and culture. It will make great contributions to the establishment of civilized cities and a harmonious society."

Law of Adaptation has great enlightening significance on the ideological and political education in the university. University students will often be confused about their life, thein family and the society in their study, work and daily life. Knowing the theory of law of adaptation is helpful for guiding teenagers' practical behaviors. It helps students adapt to history, universe and harmony, and guides them to contribute to the construction of the harmonious society. *Law of Adaptation* tells us the essence of traditional Chinese culture, spreading the positive energy. It is the culture of Yantai, the culture of our nation and the culture of the world.

In the era of globalization, the Chinese government is also vigorously implementing the strategy for culture export under the situation of world multipolarization and cultural diversity, which sets off a new upsurge of translation of Chinese classics. It is a perfect way to let China be known better by the world and be part of the extensive global communication of culture. Just as Professor Zhuo Zhenying, the convener of the discipline committee of English translation of Chinese classics once said, "English translation of Chinese classics has a great practical significance in the promotion of national culture and cultural reconciliation of the east and the west, and the maintenance of the native cultural identity." Ji Xianlin[1] also said, "We need not only the open-minded bringism (bringing from foreigners) but also bold export (sending to foreigners) to transform the focus of translation from English-Chinese translation to Chinese-English translation." Therefore, translating *Law of Adaptation* and introducing its theory to people from all over the world is a beneficial attempt that adapts to the idea of law of adaptation to follow the trend of the times, broadcast Chinese culture and harmonize people all over the world.

It took more than 2 years to complete the translation of *Law of Adaptation* after I was entrusted to translate it. During the time, I have benefited a lot from the vigorous support of many students, colleagues, friends and the leaders of School of Foreign Languages in Ludong University, especially the Provost of University of Bridgeport, Doctor Stephen F. Healey, who had given instructions on part of the translation when I visited there in 2013. As one of the research achievements of "Students' International Affairs Research Studio", the translation also got special financial support. Although I have finished the translation, I have been keeping modifying. Mr. Xu Jun (1954-) once said, "Translation is not only a personal petty thing about whether the translation is good or not, it is more about the inheritance, communication and development of culture." Therefore, I was still uneasy when the translation was about to be published. The source-language writer quoted lots of allusions, proverbs, ancient poetry and even dialects, and put forward unique philosophical theory and view point, which are very difficult to translate. In order to solve those problems, I have taken the measure of translator's annotation and made reference to many books, periodicals and other materials on Internet. It is a regret that I cannot list all of them here. Meanwhile, I have also made necessary communications with the source text writer from time to time. Despite repeated polishment and consideration, the translation cannot in the nature of things be free from errors. If really so, please don't hesitate to correct us.

Han Shuhong　Wu Yuanqing

Sept. 17th, 2015

---

[1] Ji Xianlin (1911—2009) is a famous historian translator and writer.

Preface
—In the name of daughter

Write a preface in the name of daughter with reverence and awe.

Juniors scarcely write a preface, especially daughter write it for her father. This book is a work with my father's efforts and experiences. As a child, I should not comment anything, but my father asked me to do so. I know some experience in the book well, and blood relationship passes from generation to generation. Therefore, it is proper for me to write a preface, hoping to throw away a brick in order to get a gem and serve as a modest spur to induce someone to come forward with his valuable contributions.

My father took enormous efforts to write this book, which has cost several years from the beginning to the end. Whenever he had time, he worked hard to summarize his experience and write down his comprehension. Reading this book, I am deeply touched. Everyone looks at the world with his own eyes to find life changes and inspiration. This is so-called enlightenment. Sometimes it is an idea that suddenly comes to mind or it may become a perceptual thing. However, it is no easy to write down the idea or the perceptual knowledge in words, to sublimate them into theories or logos and make them come down.

This book writes enlightenment beginning with life. Everyone's life is like a small boat in the time river. Everyone pursues wealth with different understandings of wealth. No matter how different the understandings are, there is an unchangeable standard about the purpose of pursuing wealth, which is to get happiness. Happiness is not only the purpose of wealth but also the wealth most difficult to get. This book writes about these in detail and tells you how to get the precious wealth from every detail.

This book is full of wisdom and philosophy of life. Our biggest enemy in life is ourselves. Therefore, only when we have a complete and deep realization of human nature, can we fulfill our potential most and obtain the greatest success. These ideas are well known to any educated person, but they do not always achieve them. The key is that our circumstances are different from each other. Factors that can help one get success not necessarily make others get the same result. Therefore, anyone who wants to get success should grasp the nature of things. That's why this book will discuss human nature and ways of life alone. "No fish can survive if the water is too clean. For the same reason, no company can bear one whose requirements are too critical." "We shouldn't use perfect standard to measure others' emotions and our feelings." The wisdom of life and methods of realization will touch readers. After reading the unique and wise words in this aspect, readers can find ways suitable to themselves.

Family and career are two resources that make people feel happy and influence the society most. Doing these two things well has a great influence on people and society. If we are successful, we can advance or retreat so that we stay in an invincible position. If we are unsuccessful, we cannot advance or retreat so that we stay in a hopeless situation. My father has a deep realization with many years of thinking and experiencing. This book is written in pearl-like words from the bottom of my father's heart, which is like the irises and orchids in the room and worthy of tasting.

The later part of this book is the sublimation of my father's thought, which is the essence of the whole book. The enlightenment, inspiration and discipline of life are all in the book, which does not need any of my superficial comments. The wise readers will see their wisdom.

The preface comes to the end. I am afraid that I cannot express the idea in this book, feeling that everything goes as my father said in the beginning. "Time is just like a river, and life is just like a boat. Light is needed in the dark, bravery is needed in the torrent, guidance is needed in the confusion while we are sailing. We who are operating the boat just consider obtaining knowledge as illumination, consider accumulating experiences as an oar, and consider poring over this passage as the key of solving riddles and acquiring the truth!"

Wang Fei

January, 2008

Contents

Chapter 1 Life

Time is just like a river, and life is just like a boat. Light is needed in the dark, bravery is needed in the torrent, guidance is needed in the confusion while we are sailing. We who are operating the boat just consider obtaining knowledge as illumination, consider accumulating experiences as an oar, and consider poring over this passage as the key of solving riddles and acquiring the truth!

1.1 Living with outer coverings

Attire has characteristics of a certain era and ethnic groups all the time. From ancient times to modern times, from domestic areas to oversea areas, experts will know which dynasty and ethnic group the person comes from by learning his clothing style. Nowadays, there are a large number of professional wearings, such as customs uniforms, tax department uniforms, police uniforms, military uniforms, and so on, so that we can know their career as soon as we know what kind of uniforms they wear.

One's appearance can be the reflection of his immanence. Thus a fortune-teller tells fortunes by one's face. A doctor uses 4 methods of diagnosis, namely to look, to listen, to question, and to feel the pulse, among which to look is to watch patients' face just because we can see one's ideas, conditions, knowledge etc. by learning his face. Temperament can show one's immanence, just as Su Shi[①], a great poet, said in his poem that "Reading makes a man graceful." Applying this more widely, we can see one's innermost being as well as his education, personality, habits and spirit, even his family issues, the conditions of

① Su Shi, a poet and litterateur of the Northern Song Dynasty.

his country, and his hometown's traditions from his appearance.

Therefore, human beings are all living with outer coverings, which can be the surroundings or the inner world. The covering is always on his face no matter it is good or bad, no matter how high the civilization level is. The covering can be divided into national part, urban part, working part, inner part, and so on. It can be reflected by one's appearance that whether one's country is puissant or not, whether his city is civilized or not, whether his work is smooth or not, whether his family life is happy or not, and whether the person is open-minded or not. There is no need to inquire about the issues since we can clarify everything by reading one's face.

Surroundings can be divided into 2 parts: inner surroundings and objective surroundings. Surroundings consist of space, the elements of the space, and the relationships between the elements. Different people in the same family, the same working place, and the same society can still be in different surroundings. The surrounding that one is in is a unity of the inner one and the objective one. One's outer covering is made of objective elements used as materials and inner world used as bond.

1.2 Fortune is in your sight

1.2.1 The definition of fortune

"Human beings die in pursuit of wealth, and birds die in pursuit of food." One can take pursuing the fortune as the priority in his life. However, fortune doesn't only mean money. It is a general term of all kinds of richness. Money can be a kind of fortune. Similarly, health, friendship, posterity, happiness, knowledge, intelligence, tolerance, kindness, helpfulness, vivacity, smile, harmony, diligence, unity, and so on all belong to fortune. Civilized things that do good to people's mental and physical health can be called fortune. Fortune is one's possession that makes him happy, while happiness is the kind of fortune that people are most eager for and that is most difficult to own. All fortunes can be considered as methods. But only this kind of fortune—happiness—is our purpose of life. Happiness is the most valuable fortune in the world.

One should pursue more fortune and more varieties of fortune in his life, and that makes a perfect life. One may not be happy when he only has money, because he lacks other kinds of fortune. Although the saying that "I am so poor that I only have money" is just a joke, it indicates that only having money does not mean happiness.

1.2.2 Fortune is in front of your eyes

Having a grasp of mundane affairs can be thought as treasure, and being experienced in the way of the world can be considered as fortune. Fortune exists everywhere and can be attained easily anytime you want, but people usually seek far and wide for what lies close at hand.

Someone has no ideas about cooking and abandons the food if there is something left, forgetting that "Diligence is money spinner and austerity is cornucopia."

Someone hardly pays attention to the science of diet, having no ideas of matching the nutrition, or recklessly stuffing himself with food, in which way eating all kinds of nutritional food is still in vain. Or someone does not have breakfast, or starve himself for a long time; as a result, he gets some diseases like gallstone or hypoglycemia, not knowing that 80 percent of the diseases are related to eating.

Someone does not know that all kinds of recipe have something to do with science. Improper action can not only damage the nutrition, but also produce harmful elements. So it is not strange for those people who are used to cooking in the wrong way to obtain less nutrition in his life.

Someone does shopping too casually, never caring about the clothing matching, and always feels regretful about the clothes bought, putting them aside or throwing them away after wearing them for several times. He always buys clothes, but never has clothes to wear, forgetting the old saying that "One should spend three lifetimes learning wearing and five lifetimes learning eating". We can't wear or eat too casually! It may not be too much to give others a lot of money, but it is a kind of crime to waste things.

Someone looks down upon the waste and abandons things, throwing them at random, never realizing that trashy things can be sold to gain fortune after accumulating them.

Someone gets up late in the morning and has no physical exercises in the evening, not knowing that fresh air in the morning, outdoor activities in the evening, and physical exercises all have a better effect on one's health than a lot of nutritional things, and therefore, they are fortunes, too.

Someone only knows doing physical exercises, but pays no attention to protecting his health. One should know that one's health is based 30% on physical exercise and 70% on taking good care of it. It is more troublesome to use our fragile bodies at random and then heal them painfully.

Someone is always upset, seeming like that the doom's day is coming soon, never knowing that it is a kind of treasure to have a good mood, and that it is a kind of fortune that one can own however much he would like to have.

Someone is crazy about making money when he is young, and uses money to keep life long when he is old. Working all day long is only for money, eventually seeing doctors to heal his disease by spending all his money. He never knows that health is the greatest fortune for starting his own business.

Someone becomes anxious, arguing or even fighting with others and not cooperating with others in order to make money. Finally, he delays working, wastes time and fails to make money, forgetting that the greatest fortune is owning a happy family life and that if the family lives in harmony, all affairs will prosper.

Someone makes friends or business talks outside, being modest to others, knowing that harmony brings wealth. But he puts on airs when he mixes socially with his neighbors, giving up the ready accessible relationship, and forgetting that "A distant relative is not as good as a near neighbor."

Someone never uses his own ideas, and dare not be his own boss so as to create the opportunity to make great fortune. On the contrary, he begs others around to be in charge of him, and makes fortunes for others, only earning a little money for himself. One should know that daring to think it over and put it into action is the first step to success.

Someone is not good at mutual communication, pretending to be lofty and being disrespectful. He is only enthusiastic about contacting with his so-called benefiting partners, not knowing that human relations are just the source of fortune.

Someone is acrimonious, not allowing others to have disagreements with him about anything, not knowing that tolerance is also a kind of fortune. When you ask others to do something, especially when it is the first time for him to do it, it does not matter if there is something not satisfying you. Just let him do it, especially on the condition that there would not be any bad results even if he did something wrong. One shouldn't only pay attention to the accomplishment of some issues, but give others a chance to train themselves when the issue must be done but it has no influence on any other things even if it is completed imperfectly. It is a gift we send to him because we help him to finish it better after he does it in a wrong way. In this sense, people who have these ideas and put them into practice consciously must have a great amount of both spiritual and material fortunes, because they can change the accessible things into gifts and send them to others anytime they want it, so that they can also gain others' reciprocation, although maybe they don't mean to it.

Someone is very intelligent, creative, experienced and has numerous new ideas, but he never considers the merits as fortune, not collecting and arranging them. Someone has met up with a lot of issues and people, but fails to realize that it is a kind of fortune, considering that only money can be called fortune, not knowing that thoughts and ideas are more valuable for life and the whole society, and that "A masterpiece may be eternal while officer career can be nothing." Young people are intelligent, practicing as well as feeling the life; adults are in the season of acquisition. No matter what seeds we put in the soil and whether the fruits we get are sweet or bitter, they are all fortune that deserves acquiring carefully. We should treat our experiences as experiments, analyzing and concluding them carefully so as to summarize something useful, which will be good for us to educate ourselves as well as others. The appearance of an inspiration or an idea may never happen to you again during your lifetime. Therefore, once it appears, we should seize it and value it, producing a new idea or a new piece of regulation.

Every person has his own usefulness no matter he is good or bad. We can really obtain something no matter what we do. Any issue can be considered as an opportunity no matter it is good or bad. They are all resources no matter the things are needed or abandoned. Diligence, knowledge, and intelligence are fortune themselves; laziness, ignorance, and fatuity are poverty themselves.

Fortune needs creating, needs recognizing, and needs using in a proper and reasonable way so that it can be considered as the real fortune. Everything around us can be thought as fortune if we do them well.

1.3 Worried looking comes with bad luck while smiling face comes with good luck

Fortune can be divided into 2 parts: material fortune and spiritual fortune, which is known to us all. However, mood fortune is usually ignored among the spiritual fortune. Everyone has mood fortune, and someone realizes while someone doesn't. We can live a happy life with a good mood while life can be a disaster when we are in a bad mood.

Good mood is health, happiness, and fortune. Is there anything more valuable than good mood? No, nothing. Human beings are living in a certain mood. Materials can help produce a good or bad mood while human beings themselves, which are the essential and primary cradle of the mood, can also create a good or bad mood. One can own material fortune but may not have a good mood. Similarly, one does not have material fortune but may still have a good mood. Whether we have a good mood or not depends on ourselves, not on the surroundings. A good mood is created by ourselves but not given by others, and is a kind of fortune that we can own as much as possible. Lincoln said that happiness is controlled by us. Fan Zhongyan[①] said that "One should not be pleased by external gains, not be saddened by personal losses." The nervous system can maintain a good condition, the brain and the hormonal system can cooperate with each other perfectly when we have a good mood. Then our bodies can produce all kinds of beneficial materials, so that the whole body is in a good condition. It is more important for the body to adjust itself on the basis of a good mood. Good mood and health are complementary to each other. The best doctor is a good mood.

Having a bad mood is a kind of disease, pain and loss. All the organs will be in a mess when we have a bad mood, failing to accomplish their own functions and producing harmful materials. It is a natural result to get diseases when we are in such kind of condition for a long time. Illness finds its way in by the mouth. Diseases find their way in by the mood. Death can be caused by extreme despair. A bad mood is one's own enemy. Since it is so, why do we keep the bad mood? Such people are just stupid.

Of course, even if those who see through the emptiness of the material world can still have great pressure when issues that have bad influence on their fame, fortune, or relationship happen, resulting in the people being discouraged. But the key is that clever people will shorten the suffering time as much as possible while stupid people are suffering from it for a long time.

We can figure everything out if we can truly understand the problem of

---

① Fan Zhongyan, a famous politician, litterateur, militarist and educator of the Northern Song Dynasty.

life and death. We can have a good mood if we can see through the emptiness of the world. The world is just a place where different things mix up, including satisfactory things and unsatisfactory things, things that we admire and that we can't stand. We can get a good mood if we can only take good issues into our mind. Otherwise, we will get a bad mood.

1.4 Clothing should be nearer while food should be farther

What do "nearer" and "farther" here mean? They are regulated by life sequence. For instance, animals are nearer to human beings, while plants are farther from human beings. Similarly, natural things without life are even farther.

Clothing should be nearer. Fur is the best clothing material for us to wear, then the materials made by plants, and finally materials made by some chemical substances. Animals have evolved to the present level in order to keep warm, to adapt to the environment, and to live in the world. The fur we wear is the marrow which has been in evolution for thousands of years or even longer. With a soft texture, floral material has no irritation to our body, the clothes out of which wear comfortable. Chemical fiber has an irritation or even can be harmful to our body.

Food should be farther. The most nourishing food for us is mineral composition which exists far away from us, and then is the plants, and finally is the animals. Animals enjoy the plants and absorb all of the marrow, changing it into their energy and nutrition, supporting themselves to live their life. They even can change it into part of their bodies and take use of the marrow through their physiology system. And finally, they keep the harmful substances and wastes in their body, resulting in the meat becoming the sub-quality product. The reason why animals get diseases is that there are too many harmful substances in their bodies which prevent the circulatory system from operating normally. Besides, people are nearer to animals. If we eat and digest the meat of them, it seems like that we attain their rubbish and harmful elements directly. There is an old saying that "You will get what you eat." It is possible that the blockage in the blood vessel or chronic inflammation which belongs to the animals you eat transfers to your own body. There is no doubt that it will do harm to human beings' health and shorten people's life. It has been proved by practice that people who are used to eating meat have short lives. It is true that animal meat is nutritious, yet it is a mixture of good and bad things, thus its benefit is far less than that of natural plants. As for animals, marine animals are farther than the terrestrial animals away from people, because human beings live on land. So deep-sea products are better than offshore products; offshore products are better than terrestrial animals; wild animals are better than domesticated ones. Growing up with the support of water, nutrition, sunshine in nature, plants are the essence of nature. Besides, what they can provide is just what people need. They are certainly good clean food. So we prefer to eat the marine animals rather than terrestrial animals and we prefer to eat vegetables rather than meat. Minerals are essential micro elements for human beings and the essence of nature. If using them properly, people will be energetic; if not, people will get some diseases. Therefore, people prefer mineral water.

People's experience also proves the saying that "Clothing should be nearer while food should be farther." For example, beef is not as good as mutton, mutton is not as good as chicken, chicken is not as good as fish, the meat of big fish is not as good as that of small fish and the meat of small fish is not as good as shrimp. The meat of birds is better than that of beasts. The animals with four legs are inferior to the animals with two legs and the animals with two legs are inferior to plants that have just one leg.

1.5 Start from now and it is never late

It will be never too late to take actions whether a person realizes it early or not. We realize the standard of being a man and the truth of life, and we regret what we have done and feel pity for the undone. It does not matter. From now on, do the right things to make sure that there will be no regret. It is not too late to mend the fold even after some sheep have been lost.

What we have thought does not necessarily mean what we can do. Thinking is different from doing. There is an old saying that "We can do whatever we want to as long as we can think of it." However, people always cannot match the actions with good ideas. Since people's words have their own customariness, it is always the case that people have realized something in the mind but repeat the mistakes in daily life. It needs lots of effort to make sense of something, so we had better make efforts to do as what we have realized and learn to be wise. It will never be too late, so just start from now.

Fate is under our control. We have the absolute rights to choose how to lead a life and make full use of all kinds of resources. We have the absolute rights to educate our children. We should make plans for our own life. When we are born, we are as pure as a piece of paper. We take knowledge and actions as pens to draw the picture of our life. God will help those who are faithful. People will help and respect those who are responsible for and respect themselves. People love persons who love themselves. When you are trapped in difficulties, the best way to get out is rescuing yourselves. Men should have guns during war times and should make enough money during peace times. From now on, we should create the new life and it will never be too late to start.

There is a famous story.

> Hui Zi① attended the funeral of Zhuang Zi's② wife. Zhuang Zi was squatting on the ground and beating a container, singing. Hui Zi said to Zhuang Zi, "Your wife has been with you for so many years and has given birth to babies until her death. When she is dead, you should not beat the container singing instead of crying. It is totally wrong." "No, you are wrong." Zhuang Zi replied, "At first, I really feel very sad, but after discreet thinking, I understand that she had never been born. There is no element forming the physical shape of her. In the seemingly world, gas changes into figure and figure changes into life. Now it changes back to death. This circle of life is just like season alternation. My wife has been dead and placed in the universe peacefully, so there is no sense in crying for her."

Zhuang Zi is the only person who can be so open-minded. Common people won't be brave enough to do like this. To be the first one can always be remembered.

Life will be excellent because of making sense, and victory will be splendid because of actions. We need bravery to start from now and we need bravery and perseverance to realize our ideas. One weakness of personality is that people have difficulties in defeating themselves. If the dragon can't fly in the sky, it will always be a worm. Take actions bravely and start from now. It will never be late.

---

① Hui Zi, a famous politician and philosopher of the Warring States Period.

② Zhuang Zi, a famous ideologist, philosopher and litterateur of the Warring States Period. He is the representative of the Taoist school.

1.6 Thanks giving

We are not necessarily obedient to our parents, yet we must be brought up by them. We do not necessarily make contributions to the world, yet we must take up the resources and consume them. We are not necessarily helpful to our friends, yet we must have received friendship. We do not necessarily devote ourselves to our beloved, yet we must have enjoyed the warmth of the family. Since we have received so much, we can never say that we suffer losses or we have made contributions too much. We are all passers and receivers, so we all should give thanks. The tree of loving needs thanks giving. A heart with thanks can produce much more thanks, if not, thanks will disappear. An English saying says: Love is the mother of love.

We are members of a family, so we should contribute to the family. We are civics of a country, so we should contribute to the country. We are one of human beings, so we should contribute to human beings. Since we are benefited in this group and we have a place, our contribution should not be absent.

1.7 Don't good people live long?

Good people don't live long. Good people burden a lot. Even when it is a very casual commitment and the person who hears it doesn't take it seriously, the good people will have to force themselves to perform it. Good people don't take good care of themselves, always being considerate to others, and they don't have enough physical and mental rest. Good people never trouble others. As time passes, they are ignored, not cared, or even forgotten by others. When they need necessary help, no one can help them. Good people defend other people less, but are easy to be preyed upon. Good people never release themselves, never relax themselves; they have too much thought burden, too much pressure and too much grievance, yet they will never tell others. Good people exercise self-discipline strictly, and never transgress the bounds. Good people save face, making themselves aggrieved in order to achieve others' aim. Good people can't stand others' temporary apparent sadness and give happiness to others, leaving sorrow to themselves. Eventually, good people hurt themselves early; they have nothing to do with longevity. As the saying says, "Good people don't live long, while bad people live longer."

Although it contains the pray that good people live longer, the bad people live shorter, but actually it is hard for good people to live longer because of the heavy burden.

Good people live long. From ancient times to the present, both at home and abroad, lots of people with longevity summarize that being a good person is the secret of longevity. Good people are kind people, and doing good has lots of benefit. Everyone has mercy, and doing good can make them comfortable. Doing good can produce the feeling of superiority, the feeling of accomplishment and the feeling of being wealthy, which not only is good for the psychological balance and can buffer the shock of social injustice, but also is good for physical and mental health. Because man is a spiritual animal, spirit is the master of the body and the main content of life. If people have a high spirit, they will have a good health. The highest goal of man's life is to realize himself; the last goal is to get the peace in his heart. Doing good things and being a good person will make people feel calm and happy, and the mechanism of their bodies will be operated to the best state in the pleasant environment. As a result, they could sleep well, eat well; their body will be good naturally, and longevity is natural.

In that case, do good people live long or short on earth? This is a problem of degree: excessive paying out to others naturally does harm to one's body, while doing good deeds as much as one can is good for one's health. One's doing good deeds should be based on one's abilities, so that the good things one does can last long, and one can do more good things, which is good for oneself and society. Franklin ever said, "Keep healthy, this is not only our duty, but also a social obligation." Certainly, at the critical juncture, or at the moment of sacrificing himself, longevity isn't so important to the good person. Even though good persons can't live long, their spirit will last long. They exchange their short living for others' serenity and they use their dazzling light to brighten others' world.

1.8 On loss and gain

The knowledgeable persons can obtain things while the confused persons may lose things. Wise persons gain while stupid persons lose. The kind persons gain while the evil persons lose. The virtuous persons gain while the resentful persons lose. The sensible persons gain while the insensible persons lose. The modest persons gain while the complacent persons lose. The giving-out persons gain while the asking-for persons lose. The giving-in persons gain while the vying persons lose. The wide-minded persons gain while the narrow-minded persons lose. The happy persons gain while the upset persons lose. The frank persons gain while the sorrow persons lose. The tolerant persons gain while the intolerant persons lose. The firm persons gain while the anxious persons lose. The obedient persons gain while the intractable persons lose.

1.9 Fame, benefit and emotion

Fame involves celebrity, reputation, status, etc. Benefit involves the political or economic profit. Emotion involves family ties, friendship, love, etc.

"Fame, benefit and emotion" are the entire content of life. Every action is taken all for fame, benefit and emotion. The concrete carrier of them is material or spirit. They are complementary to each other and never apart. The highest state of them is being at ease. Fame and benefit are sometimes obedient to emotion. For example, some people love beauty instead of career. Some people die for revolution. Emotion sometimes obeys to fame and benefit. For example, princes fight for the throne; children fight for family property and status; some people exchange love for money, etc.

Emotion should be treated with the most serious attitude. The importance of fame, benefit and emotion will change at different times. However, emotion is the most complicated thing and it needs to be dealt with seriously. People are always surrounded by all kinds of emotions: marital love, comradeship-in-arms, classmate-friendship, family love, indifferent feeling, and hostility, etc. and have to make choices among these emotions. Human beings are animals with emotions and they should consider emotions first while communicating. Emotions can be light or heavy. Friendship that is based on years' connection can be broken up by several unpleasant words, a selfish action or a few losses. Emotions can be priceless or worthless. Some emotions can be destroyed by several dollars or hundreds of dollars, while other emotions can never be destroyed by any amount of money. The power of emotion can always be great, and the devotion of emotion always lasts lifetime.

"Gentleman pursues wealth in proper way." People work hard to get fame, benefit and emotion. Fame, benefit and emotion all belong to profits which are rooted in selfishness. People are born for profits and die for profits, too. The trace of human life is profit-making. "People die in pursuit of wealth; birds die in pursuit of food." "People are busy coming and going all for pursuing money." "An officer works hard to make a living, a monk works hard also for making a living." Men fight for benefits and trees fight for sunshine. The society aims to develop and a single man aims to get money. The place where there are fame, benefit and emotion will be the most struggling place with cruelty and without humanity. Pursuing benefits is mostly likely to happen to the middle-aged and the young. Thus, most of the social problems are caused by these people. Among them, there are mainstays of society, heroes who can turn the tides and models who care little about profits. In addition to these people, there are also villains, robbers and war criminals who do harm to the country. Profits consist of proper profits and improper profits. People should run after the proper profits instead of the improper ones. Sometimes proper profits are accompanied and are influenced by improper profits; sometimes improper profits can ruin people's life. Xun Zi① once said, "People who view justice over profits are honorable. People who take profits as the priority over justice are shameful."

Parents take the responsibility of fame, benefit and emotion. Parents devote themselves to bringing children up in various ways with no regret. Their motion is fame, benefit and emotion. However, they take emotion as the most important one.

Emotion: The love between parents and children is the greatest emotion in the world. Parents devote themselves without any selfish purpose and they do not expect any returns. The responsibility is overwhelmed by emotion, because responsibility is negative, while emotion is active. Emotion is holy and concentrated.

Fame: Parents will be proud and happy if their children are promising. Who will not pursue this?

Benefit: The old tradition that "people bring children up to support them when they get old" reflects the aspect of benefit. Though parents do not need children's support in the families of good conditions, yet they will be glad to accept more support from children.

Driven by fame, benefit and emotion, parents devote themselves to chil-

---

① Xun Zi, a famous ideologist and educator of the Warring States Period. He is one of the representatives of the Confucian school.

dren and children receive parental love. So, what the children can do is giving thanks and love to their parents.

Career success depends on your control of fame, benefit, and emotion. What is it that gathers people to a unit? It is benefit. A unit without benefit won't exist forever. The only way to operate a group unit successfully is to fully motivate the enthusiasm of the entire group to strive for the determined aim. Only two ways can motivate the enthusiasm of the entire group. They are motivation of profit and instruction of spirit. Profit and spirit only have three points. They are fame, benefit, and emotion. The elitists in society pay more attention to dream, responsibility, and contributions besides the general pursuit of fame, benefit, and emotion. These people are a noble representative of the spirit of mankind, and therefore their spirits are worth studying and emulating widely so as to purify the soul and enhance the quality of all humankind.

Life is to pursue harmony. Life is precious for happiness and health. People will be tired if they pursue too much on fame, benefit, and emotion.

1.10 The effect of aims

1.10.1 Aims inspire people to go forward

As long as one sets a big goal, his whole body will respond to the goal and every organ of his body and spirit will automatically adapt themselves to achieve the lofty goal. As long as one sets a big goal, all is to provide inspiration, hints, test, and help for achieving the big goal and all can be used by the goal pursuer and all is served for achieving the big goal. The way to solve problems is in the subjective and objective circumstances.

People who set a high goal suddenly feel younger, more vital and energetic, and find that everything goes well and the goals can be realized. They feel that they are filled with new energy continuing to enrich their heart and soul all day long. They feel their bodies bigger than before. In order to realize his dream of going to college, a French old man started self-study from the age of 75, finally entered university at the age of 85, and got his doctor's degree at the age of 90. On his 100th birthday, his classmates around 30 years old came to celebrate his birthday, and listened to him talking about the truth of life. He said, "Age is only a number, and it is unrelated to success!" If this old man had not had such a high goal, he probably would not have lived so long.

If one did not have a goal, he would be desperate, tired, depressed, and he would feel his body smaller than before and personality lower than before. Those people who think they are too old to get achievements do nothing but eat and wait to die. Their bodies will adapt themselves to become weak and ill, continuing to go down. Xun Zi once said, "If one has a big goal, he will ignore wealth; if one has justice in his heart, he could disdain the corruptible leaders; if one pays attention to improving his personality, he will pay less attention to material comforts." All these are the effect of goals.

In fact, spirit is the guide of the body and the body is the weapon of spirit. Different spirits will lead to different worlds. For the same body, high spirit enables him to gain success, while low spirit leads him to hell.

For a boat without directions, bad times can appear at any time; for a person without a goal, he may be abandoned by society at any time. A person who is afraid of working hard has to work hard forever; a person who is willing to endure hardship will enjoy the happiness of achieving success; a person who is indulged in material comforts cannot enjoy them forever. If one blindly seeks an easy life, he gets nothing; if one strives forward, he will surely live a comfortable life. Wealth is not the essence of happiness, and poverty is not the source of depression. A man without high goals must bear double sufferings. One's success can never surpass his objective. As an old saying goes, "Establish high aims, and you'll fall into the average; establish medium aims, and you'll fall into the inferior; establish low aims, and you'll get nothing." Expectations should not be too high, and steps should never be stopped. One should hold his ultimate goal and never give up until the last moment. As long as one strives for his objective, he will have abundant harvest.

1.10.2 Single target hinders the development

In terms of career development, one should have more than one goal which can be developed simultaneously or left for reservation. In this case, there will be less adventure and less unexpected things when one goal is not achieved.

Investing in stocks can be an example. Do not invest all the money in one stock. Instead, it is advised to invest in two or three stocks. If so, one is likely to be blessed with good luck all along. If one bets on only one stock, he will get tremendous pressure while encountering unexpected circumstance.

The products of a company should adapt to market requirement. Develop another generation of products while producing one generation of products; research another generation of products while reserving one generation of products. Only by this can the company live long.

Undoubtedly, one must advance bravely towards his decided target without hesitation. However, at the same time he is supposed to establish and take other objections into consideration because they tend to be better goals or alternative objectives. In this case, his life boat can travel smoothly.

1.11 Human's rank order

Marx said, "Human being is the sum of all social relations." A person's rank order can not be divided according to his official position or wealth. Some high-ranking officials and plutocrats can be inferiors. Human beings' rank order should be divided according to their personality, characters, the way they deal with affairs and civilization degrees because these factors reflect the relationships between the essence of human beings and society. Human beings can be divided into five groups according to these factors.

Top-class people: people being kind and graceful. They are kind-hearted, tolerant, philanthropic, merciful and loyal.

Second-class people: people being amiable. They are characters of mild manners, compassion, affection and faith and they are likely to treat people gently.

Third-class people: people being indifferent. They are composed and insipid. They are neither good nor evil, neither hostile nor friendly, neither cold-minded nor enthusiastic.

Fourth-class people: people being rebellious. They are not friendly and difficult to get along with. They are suspicious, changeable and crafty and tend to gauge the heart of a gentleman with their own mean measure.

Fifth-class people: people being hostile. They are aggressive, narrow-minded and have strong hostile feelings.

The formation of the above five groups is closely related to inheritance and surroundings. However, its decisive factor is human beings' own cultivation. Harmonious society needs more top-class and second-class people.

1.12 Human beings should live like this

Human beings are the masters of the earth. They are the earth's most intelligent and most civilized creatures. Their life should be the most civilized, most reasonable and happiest one, but human beings are far from reaching this realm now, or developing in the opposite direction. We can't say that it isn't the intelligent human's tragedy. All human beings should wake up and strive for the achievement of the ideal human life.

1.12.1 Living conditions since human appeared

(1) From the time when human being appeared, especially after privacy and slavery were created, half the population died in most extreme painfulness, fear and misery owing to contrived factors. Being treated humanly is a luxury. With scientific development, it's getting increasingly easy to commit more murders.

(2) Since human beings came into existence, there have been continuous scrambling wars ending up with a field of littering corpses and a river of blood. Scientific development and more and more fierce arms race cause increasing devastation of the war. That the most advanced weapons are used to maximize benefits is similar to gaining benefits with sticks and the jungle law in animals' world. There appears terrorism in recent years, which adds terror and insecurity to human living.

(3) Since human beings came into existence, they have seriously exploited natural resources, expanded territory to the South and North Poles, even into every corner in the universe, which has damaged the living environment and resulted in the increase of natural disasters.

(4) Since human beings came into existence, about one fifth of population have been living and working under the hardest and worst environment, owing to the great difference between the rich and the poor, and to the bullying and exploitation. There are still people who are unsheltered and hungry.

(5) Conflicts in ideology, politics, economy, military and other aspects generally exist among countries and bullying and power politics happen constantly.

(6) Bad phenomena such as severe depriving, framing, deceiving, robbing and murdering exist in the society. Unfair and unjust events happen eve-

rywhere.

(7) Since privacy came into existence, corruption within groups and countries has never stopped, instead it becomes increasingly serious.

(8) Serious technology monopoly and trade protection impose numerous barriers to scientific development of the world and harmonious life.

(9) Domestic violence and troubles do harm to the living quality of mankind. Domestic privacy has great harm to stable conditions of the whole society.

(10) Most people live with a heavy and imbalanced heart, therefore, pleasure and happiness seem a luxury to them. Many people have flattered others, pretended to do some things and lost themselves for their own benefits.

This living style is the most selfish, most pathetic and most vulgar one and it is similar to animals' living style. Humanity has been far from entering civilization. Such living style leaves no choice but to quicken the devastation of mankind.

Why do human beings form countries and develop alliances to compete for own benefits? Humans are not supposed to do so and it is the behavior of the beasts. Do not wait for the time when countries naturally disappear because it is a long way to go. If the countries could really disappear one day, the earth would have already been broken. So far mankind in fact has the awareness, consensus and ability to change regional division and deformed life style.

Everything in the universe adheres to the rules of survival, change and death. Since mankind dies sooner or later, why don't they have peaceful coexistence when they are alive and have the chance? Is it necessary to claim hegemonies? Since everyone must end up in death, why can't they strive for common happiness? Is it necessary to fight to death? Honestly, the word "selfishness" is responsible.

There is a thought-provoking story: the prisoners were escorted by a group of nobles in the ocean, and the nobles did not give enough food and water to the prisoners, neither let them come up to the deck. Prisoners began to battle with the nobles. They destroyed the vessel, and caused the ship to leak. All the people in the ship died eventually.

The vessel is like our globe. Human beings are in the same boat and the artificial classification creates conflicts among them, causing them to fight with each other, making them get together to destroy the earth, and the final result may be that all of us will die.

1.12.2 The ideal mode of human living

As a wise master of the world, human should have a harmonious and happy life. It should be a blueprint like this:

(1) All the wars should be wiped out so that human beings can live in peaceful environment.

(2) Everyone has the complete freedom of faith and action; everyone should make a positive contribution to the society.

(3) Countries disappear in the world. Human beings live in a regional group or organization. The sole purpose of groups and organizations is to safeguard the group security and to contribute better to mankind.

(4) Industrial and agricultural productions are distributed according to the world's natural resources, and the allocation of materials should be done in a unified way without any waste.

(5) The world develops in a planned, orderly and steady way. People who waste resources are going to be punished.

(6) Everyone has his or her job and tries his or her best to do the job well.

(7) The family privacy disappears and all the properties are owned by human beings together.

(8) Products are distributed in an equal way according to one's needs. The wealth is distributed averagely all over the world.

(9) There is no competition for profits, honors, etc. All the people respect each other.

(10) Everyone is always cheerful and everyone's life is always happy.

This kind of social situation and life style sounds like a fairy tale and seems to be utopian, but in fact it is achievable. A better society will be implemented in a better imagination, as long as the imagination can be implemented. So how does that happen? The only way is to eliminate human selfishness.

1.13 Everyone is inseparable from philosophy

It is widely acknowledged that practice is the source of theory and theory guides practice. But when mentioning theory, some people view it as the most profound thing—a profession only for the theoreticians. In fact, people act under the direction of theory consciously or unconsciously. Whatever the result is, good or bad, once one finishes a task, he will gain experience. Gradually, such experience or experience accumulation will produce a logic concept and during this process we draw experience or lesson from it so as to guide our words and deeds later. In effect, some people neglect the importance of summarizing and naturally they will make the same mistake next time. The successful men need three kinds of experience, namely, their own experience, experience from forerunners and experience from books. Among the three ones, the experience accumulated by themselves is the most practical.

Philosophy is the highly summarized theory, the guiding ideology of life and the knowledge which everyone needs. Full of generality and abstractness, the connotation of philosophy is fathomless. The expression form of philosophy should be easy to understand, and can be mastered and used by the majority of the people. Smart people should use their own understanding of philosophy to guide their behaviors. The existence and development of things have laws and the key to know the laws is to use the laws.

Thought guides behavior. Whatever kind of thought you have, you will have behaviors that correspond to that kind of thought. Therefore, think first and then act. We should think comprehensively before we take actions, so that we can yield twice the result with half the effort and improve the rate of success.

The world consists of the independent objects and the relationships among them. For example, a family consists of family members and the relationships among them. The number of family members is determinate, and the relationships among them should also have certain laws. Therefore, the key to maintain a good family is to assure the relationships among the members, to find the correct location of every member and seize the elasticity of the relationships. Make sure that each person is relaxed, harmonious and not unorganized, and that each of them has centripetal force and his own freedom.

Everything in the universe exists and changes in the state of balance and unbalance and balance includes positive balance and negative balance. Sometimes, balance develops into unbalance, or unbalance develops into a new balance. Therefore, seeking a positive balance and breaking a negative balance for development are necessary when people are dealing with things and relationships.

Structure determines everything. While analyzing and solving a problem, we should start from its structure. Changing structure can change things, and positive and beneficial change is innovation.

Effect won't disappear. We should create positive effect on our own initiative and make our life valuable.

Everyone is a philosopher. Philosophy is the view about the world and the knowledge about the world outlook; everyone has his own opinions about the world and also has his own world outlook, therefore, everyone has philosophic thinking. Philosophers can sum up their philosophic thinking systematically. If an ordinary person can do this, he could also become a philosopher. Of course, if you want to be an "expert", you need to have advanced theories; ordinary people can also create advanced theories. If a kind of thought wants to be an ideology that could guide people's behavior, it should have two qualifications at least: one is that the thinking itself should have guiding connotation, the other is that it should be widely acknowledged. The former is self effect, and the latter is the cognitive effect. The two effects combined together will become the guiding principle. A farmer who has not attended school can also have philosophical thinking. If we set up a "research association" to research him, we could also excavate deep thoughts. If these thoughts are widely acknowledged, he will become a celebrity and his thoughts could have guiding effect. Therefore, don't be afraid of great men. Don't be afraid of saints, either. Instead, we should dare to excavate and propagate our own thoughts.

1.14 Everyone could be a saint, as long as he is aggressive

Everyone could be a saint, but not everyone can form his own words. After reading saints' books, we all agree on the saints' points. This is the cognizance on the basis of self-experience about life and the world. In another word, everyone knows it, but they haven't got down to summarize it and written it down.

Everyone could be a saint, but not everyone purifies his words. Saints' words are incisive. And their attitudes towards life can stand up to the test of history. Saints' words cannot come into being without the great awakening and the repeating purification.

Everyone could be a saint, if only everyone does solid work. Saints are extraordinary people, which is not only reflected in their behavior but also in the way they act. Common people know how to become a saint. But they are not determined to achieve it.

Everyone could be a saint. Saints are also human beings. Saints are often not well-known until they die. That is because their life has no difference compared with common people. However, as time goes by, the greatness of their thoughts starts to be noticed by people. Obviously, not all saints' words are right, and not all their words suit for each dynasty and for everyone. Cognition can never surpass reality. All truths are close to reality, and every truth has the space to be further developed.

It is good if you want to be famous. Emulating those better than ourselves is a good quality. If all the people could become saints, the world would change to the best.

1.15 The destination of human life

Choosing the way of life is choosing the destination of it.

All roads lead to Rome, and the roads of Rome lead to everywhere. The destination of one's life is determined when he first chooses his way of living.

Socrates once told a story.

> There were two women. One was Atere. The other was Cagia. They were both beautiful and young. And they were classmates from childhood on. They both had happy family and received high education. Life was beautiful and they were both happy. However, life was changed after they chose different ways of living. Atere became a nun. She prayed for people in the church. She sent the money donated by people to the poor. She was happy with what she was doing. How about Cagia? She became a prostitute. She did not lack money and happiness at that time. She experienced all the enjoyments. She was also happy. Later, they both became old. Atere was regarded as godness. And people all respected her. However, Cagia was poor and she could only beg in the street. Even when she died, no one took part in her funeral. Socrates said, "The two women chose their own ways of living. One led to happiness and the other to evilness."

1.16 We can only judge a person after his death

People always struggle for the future while they have are alive. If they make mistakes or even fail near the end of their life, all the efforts they have made before become useless. On the contrary, a common life with a sound ending is also successful.

Later generations may judge a person after his death according to his whole life. Merit is merit and demerit is demerit. We will not ignore one's merit just because of the mistakes he made in his later years. In the same way, we also won't evaluate a person just by his merit achieved in his later years.

The story of Liu Bang[①] and Xiang Yu[②] is well-known in history. They were opponents. At first, Xiang Yu had more advantages than Liu Bang. But he committed suicide while Liu won the victory in the end. They fought for so long time. The ending turned to be that Liu Bang was the winner and Xiang Yu was the loser. Many people have made comments on them. Xiang Yu was the son of a famous general. He contributed a lot to the overthrow of the Qin Dynasty. But because of his misuse of talents and also some other reasons, he failed at last. Liu Bang loved drinking and lusted for women. But he knew how to make use of talents. And he finally succeeded in his life. Although this judgment is not all-sided, it does reflect the way how people judge a person.

In the Qin Dynasty, Li Si[③] was responsible for making the law and unifying the characters during the union of the other six states and made great contributions. In order to reward Li Si, Qin Shihuang[④] let his daughter marry Li's son. However, in the year 210 B. C., Qin Shihuang suddenly died of illness when he was out. At that time Zhao Gao[⑤] pushed Li Si into the political intrigue. Enticed by Zhao Gao, they blocked the news of the death of Qin Shihuang, and spread the false lord's order. And they forced Prince Fu Su[⑥] to commit suicide and made the fatuous Hu Hai[⑦] the next Emperor. However,

---

① Liu Bang(256 B. C.—195 B. C.), courtesy name Ji, also known as Emperor Gaozu, is the founder and the first emperor of the Western Han Dynasty(206 B. C.—24 A. D.).

② Xiang Yu(232 B. C.—202 B. C.), courtesy name Yu, also named Xiang Ji, is a prominent military leader and political figure in the late Qin Dynasty.

③ Li Si(280 B. C.—208 B. C.) is a Chinese politician of the Qin Dynasty.

④ Qin Shihuang(259 B. C.—210 B. C.), personal name Ying Zheng, is the King of the state of Qin (246 B. C.—221 B. C.) who conquered all other Warring States and united China in 221 B. C.

⑤ Zhao Gao(? —207 B. C.) is a Chinese politician of the news of the Qin Dynasty.

⑥ Fu Su (? —210 B. C.) is the eldest son and heir apparent of Qin Shihuang, the first emperor of the Qin Dynasty.

⑦ Hu Hai (230 B. C.—207 B. C.), regnal name the Second Emperor, is the son of Qin Shihuang and the second emperor of the Qin Dynasty. He ruled from 210 B. C. to 207 B. C.

two years later, Li Si was killed by Hu Hai under the instigation of Zhao Gao. Li Si failed in the last part of his life. But people still regard him as the famous statesman, writer and calligrapher of the Qin Dynasty.

In the society of market economy, a "58-year-old" phenomenon occurs. Some privileged officials have been working heart and soul but they are easy to make mistakes near the retire age of 58. They want to make more money through illegal ways before they lose the power. In this way they have ruined themselves. They end up in failure. Some other people have not achieved a lot before their retirement. But after retirement they start a new business, for example, they concentrate on writing, painting, and later they become world's famous persons. What's more, some people are not even known by others when they are alive. But they become great people after they die just for what they have done in the whole life. Cao Xueqin① lived a poor life and was not well-known to others when he was alive, but he became famous for his book *A Dream of Red Mansions* after his death.

People can never understand life before the end of it. And the one who can understand his life must be a wise man. Young people cannot understand old people's behaviors sometimes. But later they all understand because they have behaved in the same way. Therefore, young people should try to understand the old and do not easily deny others since you are not old enough to judge life. You will never know it until you experience it.

Jiang Ziya② became the prime minister at the age of 80. As the saying goes, "Great minds mature slowly." On one hand, it means some people become famous at an old age. On the other hand, it meens to become great talent, you need to practice a lot.

To conclude, only after his death can we judge a person. It's never too late for a person to make an effort while he is still alive. As long as he is alive, one should obey the rules of the society. The most important thing of living is to make life meaningful.

---

① Cao Xueqin(1715 or 1724—1763 or 1764) is a Chinese writer during the Qing Dynasty. He is best known as the author of *A Dream of Red Mansions*, one of the Four Great Classical Novels of Chinese literature.

② Lü Shang(11th century B. C.), commonly known as Jiang Ziya or Jiang Shang, is an ancient Chinese military strategist who helped King Wu and King Wu of Zhou overthrow the Shang Dynasty.

1.17 Make up the lessons which one has missed

One must make up the lessons that he has missed. Sometimes one gets some experience after great cost, but that is too late. For example: (1) We don't know how to give presents. Courtesy demands reciprocity. Sending presents to each other is one of the social activities. But sometimes when we ask someone for help, we do not know how to send presents and what gift is suitable. If we send some small gifts, he may think we are not polite or look down upon him. Or if we send a wrong gift, he may misunderstand us or laugh at us. (2) We don't know how to visit patients. We need to visit patients at proper time with proper gifts. If we bring wrong gifts or make a visit in the wrong time, the result won't be good. (3) We don't appreciate what we have owned. Our environment and the benefits it offers are evidently good, but we don't treasure or don't know how to treasure them. We cannot realize their value and importance until we have lost them. (4) We are not good at communication. There is a very reasonable argument, "In modern society, 70% of success depends on social relationships and 30% depends on one's knowledge." The social relationship does not mean the negative relationship. It refers more frequently to the ability to adapt oneself to the society. Unfortunately, we do not comprehend the truth and do not know how to communicate with others. As a result, we cannot have a bright life because of lacking information and opportunities as well as others' aid. For another example, we make too many friends, but we don't know how to make true friends. We do not know traditional code of conduct among human beings and we know little about the scientific knowledge, etc. These courses that we haven't learned in advance may cause us great loss.

The development of things is cyclical and spiral. Life is short, but also long. There are a lot of rotations in a person's life: a rotation of four seasons in a year, a rotation of fortune and misfortune, a rotation of honor and disgrace, a rotation of gains and losses, a rotation of being students for several times, a rotation of changing several jobs, etc. There may also be several marriages in one's life. There are many things that we will do for several times. Life requires us to make up the classes which we have missed. The things we haven't encountered now will happen sooner or later. We always have the opportunity to face them, even after paying the cost. Therefore, people should read and study more as soon as possible so as to learn more about the society and to have more contact with intellects. Whatever problem we meet, we should seek the best solution even though the problem has been solved, because we may meet the same problem again.

1.18 Life is a ship

Life is a ship which sails in the long life river. People of high social status are big ships. People of low social status are small ships. Sometimes the ship is sailing in a calm sea and sometimes in a stormy sea, sometimes in a clear river and sometimes in a muddy river, sometimes in the deep water port and sometimes in the hidden shoals and dangerous reefs, sometimes in smooth wind and sometimes against the wind, sometimes into a wide water area and sometimes in a narrow water field, sometimes on a straight line and sometimes on detours. This ship exhausts a lot of resources and dumps a lot of rubbishes. It ships a lot of harvest and bears a lot of pains.

How should we spend our lives? Confucius said, "The gentleman stands in awe of three things. He is in awe of the Decree of Heaven. He is in awe of great men. He is in awe of the words of the sages. The non-gentleman does not understand the Decree of Heaven, thus does not stand in awe of it. He treats great men with insolence and makes a mockery of the words of sages." Lao Zi① said, "The Tao way follows the nature." A poet called Du Xunhe of the Tang Dynasty said, "The Jingxi River is well-known for its dangerous rocks the in the water, but there are no persons drowned all the year because people are very careful while crossing it; however, we often hear that there are persons drowned at some plain water area where there are no dangers at all." We should always follow the tides and obey the rules so as to pilot the life ship smoothly and safely for ever.

---

① Lao Zi (Wade-Giles: Lao tzu; also Lao Tse, Laotze) is a philosopher of ancient China and is a central figure in Taoism (also spelled "Daoism"). Lao Zi literally means "Old Master" and is generally considered an honorific. Lao Zi is revered as a god in religious forms of Taoism.

1.19 Shape and appearance are the outward manifestations of self-cultivation

This topic seems to be a bit absurd at first, especially for those people who are not satisfied with their bodily form. They will jump up for criticisms. But after meditation, we will find it reasonable and having the promoting function to improve our bodily forms. As the saying goes, "Appearance springs from the heart whereas perception turns with the soul."

1.19.1 Definition of shape and appearance

Size is not shape. Size means the proportion of every part of the body, while shape refers to the overall appearance of body form. Looks are not appearance. Looks refer to the facial features while appearance is the outward manifestation of looks and the impression left to others. Size and looks result from inheritance and body development while shape and appearance develop on the basis of size and looks after we grow up, including being fat or thin, symmetrical or not, kind or evil, cold or warm, beautiful or ugly, as well as gesture, age and appearance, etc.

1.19.2 Behavior forms shape and appearance

Obesity results from eating too much and exercising too little. Eating more and exercising less are the main reasons for obesity. Good shape has direct connection with body exercise; appearance is closely related to the habits; kindness and evilness are decided by the attitudes towards life; warmness and coldness are formed in interpersonal communications; beauty and ugliness are the outlook of being kind or evil; age and appearance are a state of mentality. The above forms of shape and appearance are all the result of self-cultivation. It is said that for one time Lincoln wanted to employ a clerk. He sent one interviewee away after the first sight of him. The staff asked him why he did so without any test. Lincoln said, "He looks bad. He is supposed to be responsible for his behavior after his 20th birthday."

1.19.3 To reinforce self-cultivation is a priority

Since shape and appearance are external performance of self-cultivation, it is high time for us to reinforce our self-cultivation. Then what are the definition and standard of self-cultivation?

(1) The definition of self-cultivation

Self-cultivation refers to a person's certain level in theory, knowledge, art, thoughts, world views, life attitude, etc. Self-cultivation is the result of cultivating and correcting oneself in accordance with the universally recognized standards. Self-cultivation is the level of cognition of things and self-control.

(2) The standards of self-cultivation

On the whole, self-cultivation should include the following items:

1) Pay attention to food and drink. In modern society, eating and drinking are not only to satisfy the appetite. More importantly, we should build a healthy body through food and drink. Therefore, a person needs to know about the right diet, and manages to follow it. If a man doesn't understand scientific diet, or doesn't care about diet, can we say he has a good quality of self-cultivation? Of course, we can't. For example, some people eat hot pot dishes while the dishes are still raw, saying that they taste good and tender. But finally germs and bacteria appear in the internal organs. It does a serious harm to the body health. Can we say that these people are well-cultivated persons? Some persons eat and drink too much, getting potbellied, acting inconveniently, and being vulnerable to diseases. This can also be the performance of not being well-cultivated.

2) Wear properly. Now, clothes are not purely used to cover the bodies and keep warm, but also to reflect our internal spirit, cultural attainment, aesthetic art, and personal temperament. Clothes can directly affect our bodily form. As the saying goes, "Learning to eat needs three generations' practice, and learning to wear needs five generations' effort." There are a lot of things for us to learn in eating and wearing.

3) Sit and stand in the right way. Habits have great influence on the form of our figures. Walking like a crab with collapsing waist and slanting shoulders, sitting with sideways body, some persons seem to do whatever is ugly. As time goes by, these bad habits cause body deformation and sickness. As the saying goes, "Persons should stand like a pine, sit like a desk clock, walk like the wind and lie like a bow." To stand like a pine means that one should be erect and straight while standing. To sit like a desk clock means that one should keep steady and decorous while sitting. To walk like the wind means that one should walk fast and powerfully. To lie like a bow means to lie on the

side with the body bent to a certain degree, not lying on one's back or lying face downward. To keep the correct posture is important for youngsters' normal growth of their bodies. And it is very important for adults to keep fit and reduce diseases.

4) Exercise regularly. As the saying goes, "Life lies in movement." Without movement, the muscle will shrink, the bone will soften, the circulation system will get into a jam, the digestive system will fail to absorb, the endocrine system will be disordered, and so on. In fact, almost everyone knows the importance of physical exercise and perhaps everyone will occasionally exercise a bit, but not everyone can keep doing exercises as a habit.

5) Life should be regular. Regular life is very important for keeping physical and mental health. Some persons spend the whole time playing mahjong or cards late into night. It is obvious that their spirit is not positive and their appearance and shape are not in good condition.

6) Pay attention to health care. Health care includes physical knowledge, common disease pathogenesis, and common sense of health products' screening. Man is composed of materials and if he wants to have a healthy body, he must provide good and reasonable materials for his body and take timely and scientific care measures.

7) Cultivate oneself. Self-cultivation is the adjustments of spirit, and the spirit is the commander of physical body. The whole body will be excellent with a good mental spirit. Self-cultivation can balance mentality, reduce diseases, and make a person lead a happy life at a certain level.

8) Be kind to others. Someone frowns all day and seems that all people owe him a lot, resulting in a ferocious face. Someone is always kind to others and the result is that he has benevolent and kind countenances as if he were experiencing the spring breeze.

9) Follow the exquisite civilization etiquette. Civilization etiquette improves people's quality, temperament and of course, body forms' performance.

10) Be tolerant and optimistic. We should be cheerful and not fussy about every detail. We had better let ourselves accept things that are difficult to accept for others and laugh at them when ridiculous persons appear. As long as we do like this, we can become a kind person and a healthy person both physically and mentally.

To conclude, health, happy mood and positive attitude are all the embodiment of self-cultivation. Self-cultivation determines posture, health, mood, mentality, career, family and so on. It is the basis of being a person. As the old saying goes, "The ancients who wish to illustrate illustrious virtue throughout the kingdom, first govern well their own states. Wishing to govern well their states, they first regulate their families. Wishing to regulate their families, they first cultivate themselves. Wishing to cultivate themselves, they first rectify their hearts ... Their hearts being rectified, their personalities are cultivated. Their personalities being cultivated, their families are regulated. Their families being regulated, their states are rightly governed. Their states being rightly governed, the whole kingdom is made tranquil and happy."① Life lies in movement and health lies in self-cultivation. People live in self-cultivation and their lives depend on their self-cultivation.

1.20 Enjoy the life

Life is a process instead of a result. And perhaps there is nobody who wants the result. So what is the best way to spend our life? The answer is to appreciate and enjoy our life from the bottom of our heart, and enjoy it everywhere in every second.

During leisure time, enjoy the sight of every scenery and appreciate all the natural plants including every tree and every flower, learning connotation of beauty with love and enjoying the gifts of nature. While walking through the street or the field, relax yourself in every step and every activity, appreciating everything with joy and enjoying life as much as you like.

In the working season, we taste all the feelings of being busy, freedom, tiredness, and bitterness. Because these are scarce resources, it is impossible to let them come back once again. Life is a one-way process without retrogression.

At the moment of treating people, whether they are an acquaintance, a stranger or a relative, they all have karma with us. We should accept, protect, and respect each other and appreciate all kinds of characters with tolerance, which is also a way to make life full.

Perceive what we own attentively, including material, emotion, health, and so on, then we will really possess them. If we did not feel or taste what

① Quoted from a Chinese classical literature *The Great Learning*.

we have consciously, we would not possess them even when we are living in a fairly comfortable condition.

We earn money to live, but not live to earn. We work to live, not live to work. If you don't live well, working and earning will lose their meaning.

Enjoy your life, don't waste it; every second should not be missed. Use your love to accept the universal creations; use your laughter to face life. If you could do this, it would not be possible to have a bad mood.

Most of us have no initiative when facing the environment and work. But it is possible to enjoy actively what we meet. And that's the point of the entertainment of life and the disappearance of trouble.

1.21 How does happiness come?

One just seeks two things in his life, which are health and happiness. Health is the biggest business and happiness is the biggest success. It is your own feeling that determines your happiness and your success, but not the evaluation from others. What is happiness? Happiness is the comfortable and delightful feeling in your body and soul. It belongs to the spiritual category. In brief, a man without pain is happy. In a thorough way, a man who is alive is happy. How does happiness come? Happiness needs to be built in all aspects.

Happiness comes from labor. Labor creates history; labor casts the future. Labor is the basis of human living. And it is also the basis of happiness. Zhu Xi of the Song Dynasty has a poem, "Why is the water so clear in the dyke? For the fresh water comes from the springhead."①

Happiness is based on the material. Materials are the basis of people's survival.

Happiness comes from spiritual life. It could not be imagined that human being could live without mental nourishment. Human beings are spiritual animals. Spirit supports human's living.

Happiness comes from careful thinking about life. Happiness does not come from the God. It is not given by others. Happiness comes from thinking about life. It is the outcome of understanding. Happiness comes out of the basis of a certain material (without worrying about food and clothing).

---

① Quoted from the poem "On Reading" written by Zhu Xi.

A man being wealthy is not necessarily happy. Money takes a small proportion in happiness. Many experts have studied the relationship between money and happiness. They found without exception that the pertinence of the two is only 0.25. That is to say, the function of owning a lot of money is only 1/4 in judging whether you are happy or not. Only money can not support the palace of happiness. "Money is not the index to measure your value, nor is the protection of happiness." When a man feels satisfied with his basic necessities of life (food, clothing, residence, etc.), money's effect will be diminishing on the degree of satisfaction of happiness. Life is to pursue happiness not money. "Satisfaction is the state of being most wealthy and having no worry is the state of being most happy." ①

A man who becomes an official is not necessarily happy due to power struggles, scheme against each other, worries about personal gains and losses, etc. The officials will rack their brains and be unable to sit down or sleep at ease in order to climb to a higher place. Being privileged, they want to make more money and after making more money, they begin to worry about it. Owing to poor popularity and unfairness in dealing with things, they will be abused after their retirement. They will have a big psychological gap in their heart and feel depressed and not happy at all.

There are a thousand hamlets in a thousand persons' eyes. Some people feel satisfied with work; some people feel satisfied with having time to read books; some people feel satisfied with helping others; some people feel satisfied with having no trouble; some people feel satisfied with traveling; some people feel satisfied with the happiness of a family union; some people feel satisfied with dancing, singing and other recreational activities. No matter what you do, if you like it, just do it.

We should be broad-minded, getting a thorough understanding of life. The universe is eternal and lives are natural. An ancient Chinese poet Du Fu said, "A man seldom lives to be seventy years old." Even if you can live for 80 years, that is about only 30,000 days. No matter what kind of way you would like choosing—being happy or sad, vigorous or silent, broad-minded or narrow-minded—life is fleeting, the stars are changing their positions, the universe is endless, and things are changing all the time. Human beings are

---

① Quoted from a Chinese classical literature *Lao Zi*.

like passengers who are going to attend a dinner party. After tasting all the foods being sour, sweet, bitter, hot or salty, they will go back. View life with ordinary eyes, and do not take everything too seriously. Thorough materialists fear nothing, and they are optimistic. Therefore, why aren't we broad-minded? Do not make a rod for our own back. Do not waste our only chance to live. Wonderful life is right in front of us. Why not choose it? We should hold such kind of thought—our life is the best; our choices are right. If the way of living had been changed, it might have been much worse. Actually, that is it and it is life! We should be satisfied with the status quo, thinking our lives are much better than others'. The foundation and connotation of happiness are changing with the development of our society, but its root is the mental state. Broad mind will lead to broad eyesight. "The sea is big because all rivers run into it." "One can't show high ideals without simple living; one can't have lofty aspirations without a peaceful state of mind." My daughter Wang Fei ever said, "You feel happy because you are alive." Happiness is coming naturally without pursuing luxuries. We should be satisfied with our living. We should feel happy as long as we are alive. If we can achieve a high spiritual state like this, we are sure to lead a happy life.

Happiness is found by oneself. There is a story titled *The Happiest Farmer*.

One farmer had spent a hard time of his life and he began to feel tired about it. One day, he met a passenger who was a philosopher and he asked him sincerely,

"A man just liked me, when can I get my happiness?"

The philosopher felt so amazed. He asked, "Don't you live a happy life now?" And the farmer answered, "I must work hard in order to support the whole family. At the same time, I must take some extra money for my wife's medicine. I feel so frustrated toward my life."

The philosopher asked another question with smile, "If a man is delighted to give you fifty kilograms of gold to replace your healthy heart with a dying heart, and to replace your brain with a memory-lost brain, will you agree?"

The farmer shook his head decisively.

The philosopher said, "If someone is willing to use a luxurious building to exchange your children, and cars, good fields for your wife, will you agree?"

The farmer shook his head again.

At that time, the philosopher laughed, saying, "So, you are the happiest person in the world because you have the most precious possessions of human being. Happiness is held in your hands." The farmer released and burst out laughing. Being capable of working and paying for the one you love every day is also a kind of happiness.

Enjoy your life. Life is colorful and one can't enjoy all kinds of life styles. Everyone has his life which others can't completely experience. Don't expect to enjoy other people's life. To enjoy your own life is the best choice.

Extend happiness and shorten pain. Richard Thaler, the professor of behavioral science at the University of Chicago, has proposed four principles on the announcement of good news and bad news: a. If you have several pieces of good news to be announced, you should announce them separately; b. If you have a few pieces of bad news to be announced, you should announce them together; c. If you have a big piece of good news and a little piece of bad news, you should tell others the two news together; d. If you have a big piece of bad news and a little piece of good news, you should announce them separately. Thaler's four principles' effect is to extend happiness. "Being small and frequent" is better than "being big and scarce" for the former can bring a person more happiness. To tell a person two pieces of bad news one time can shorten the pain. Pain should be short and happiness should be long.

We are close to happiness. Happy life does not need a blockbuster career or too much money. Everyone has a happy platform. Everyone is on the platform of happiness. Whether you feel happy or not depends on how to handle happiness on the platform. Everyone's basis of happiness is different because happiness is a kind of psychological feeling; it is felt in the heart.

Happiness is different for different people. First of all, the foundation of happiness is different. Some people base their happiness on the sufferings of others, or on the injury to the public interests or even to the interests of the Nation; some people base their happiness on their hard work or on taking pleasure in helping others; some people base their happiness on the spiritual wealth, i. e. a contented mind; some people base their happiness on the material wealth or on being greedy and soliciting of advantages. Second, the forms of happiness are different. Some people think that it is happy to spend money like water or to lead a befuddled life; some people think that it is happy to play the piano, to play chess, to draw a painting or to practice calligraphy; some people think that it is happy to make a sightseeing tour. Third, different kinds of happiness will last different length of time. Happiness based on the sufferings of others, or on the injury to the public interests or even to the interests of the Nation, or on being greedy, will not last forever. Happiness based on one's hard work or on being kind to others or on a contented mind will last forever. Fourth, the prospect of happiness is different. A kind man's prospect of happiness is still happiness while an evil man's prospect of happiness must be misfortune.

How does happiness come? As long as you have a healthy body, a job to earn your own living, a beloved lover, a group of sincere friends, a grateful heart, and the wisdom to see through the world, happiness will come to you naturally. The purpose of life is to pursue happiness. All the activities of life are for the pursuit of happiness. Happiness is all around us and the key is that whether we can get hold of it.

1.22 Character determines destiny

"Thought determines actions. Actions lead to habits. Habits form character. Character determines destiny." Character is human's essence feature, is the determinant of attitude to things and the ways of behavior and is the main style of doing things. Whether one can get on well with others depends on his character. Historically, Xiang Yu was imperious, despotic, murderous and headstrong. He was doomed to failure because of his bad characters. In contrast, Liu Bang employed talents, listened to suggestions, and he was easy-going. His good characters helped him achieve great accomplishments. In human communication, if we are nice to others, they will be nice to us. If we help others, they will bring benefits to us. Some people who are affected by their characters and the environment know clearly that what they will do is bad for others and themselves, but they still do so. As a consequence, they suffer great losses. Therefore, it's lucky to own a good character.

How is character formed? Character has both genetic and acquired components. And the acquired disposition plays a decisive role. The formation of character adheres to "three-three law". Ideas in the same direction (such as they are all well-meaning) appearing three times can guide people to have the same direction's actions. Actions in the same direction appearing three times can make people have the same direction's habits. Three kinds of habits in the same direction can form the same direction's characters. All aspects of the characters determine the fate of a person.

Character needs correcting. It's hard for people to completely know their own characters' advantages and disadvantages. It's harder to change their characters. As the saying goes, "The fox may grow grey, but never good." Everyone's character has shortcomings. People must realize their shortcomings and try hard to correct them. Only after people have corrected their shortcomings can they have a better achievement in their career and get more happiness. Although it's difficult to change one's character, there are also ways to change it. If we want to change our fate, we should change our characters and we must start to do it bit by bit. Now that "thought determines actions; actions lead to habits; habits form character; character determines destiny", if one wants to change his fate, he should change his character first. If one wants to change his character, he should change his habit first. If one wants to change his habit, he should change his action first. If one wants to change his action, he should change his thought first. We can infer the "three-three law" of correcting characters from the "three-three law" of the formation of characters. That is, looking before you leap can correct the action; actions corrected in the same direction three times can correct the habit; habits corrected in the same direction three times can correct the character.

1.23 Category of knowledge

The world's knowledge can be divided into three categories according to its different functions: ideal knowledge, utilitarian knowledge and health knowledge. Ideal knowledge is the knowledge which helps foster ideals and establish noble sentiment. It includes thought, doctrine, spirit, philosophy, religion, etc. Utilitarian knowledge is the knowledge which helps people achieve high official positions and acquire wealth. It includes natural science, social science and some content of human thinking. Health knowledge is the knowledge which helps prolong the life. It includes all works, secret recipe and tips on health in history as well as attitude to life and recipe of all aspects of life.

1.23.1 Purposes and functions of knowledge

Any kind of knowledge is the summary, induction and sublimation of historical experience.

(1) Ideal knowledge

Ideal knowledge is created by people with lofty ideas for the good future and noble thoughts of human beings. It was developed through gradual perfection of generations. Ideal knowledge comes into being in hardships. We must suffer from great hardships so that we can understand things completely. The creators of ideal knowledge put forward theories on nature, world and life by standing at a detached level. The ideal knowledge with deep enlightenment and instruction function has the character of idealization and it is inherited by generations. Ideal knowledge is also the spiritual knowledge, some of which is the best hope of the progress of mankind. So the motivational mechanisms of mentality should be established.

(2) Utilitarian knowledge

Utilitarian knowledge is summarized by people through practices in order to develop economy and build wealth, livelihood and utility. It comes into being in the environment of social production; it reflects the disciplines of nature, society and human mind and it directs the economic development for people. Utilitarian knowledge has the characteristic of fast updating.

(3) Health knowledge

Health knowledge is summarized for health and longevity through the historically repeated practices by people. It offers suggestions for people to keep a good health and prevent diseases. Health knowledge is inherited from generation to generation. The content of health knowledge is relatively stable.

The foundation, preach and application of health theory for commercial purposes belong to utilitarian category.

1.23.2 The degree of people's studying and practicing different categories of knowledge

(1) Ideal knowledge. At the time when the ideal knowledge was first produced, it wasn't valued by people and even suffered taunts and taboos. In the development of history, it was not always taken so seriously. Even in the modern society, it may be not valued by people. However, it can be inherited forever. For example, the production and development of Confucius's thoughts and Mencius's thoughts are along this track.

Young people have limitless hope for ideal knowledge and are easy to accept it and stick to pursuing it . With the increase of age and life experience, people switch their focus from ideal knowledge to utilitarian knowledge. After getting old, for example, at one's fifties, he will change once again to seek ideal knowledge.

People shouldn't abandon the pursuit of becoming saints because of the failure of becoming high officials. In the long past history, few emperors could exclaim themselves as saints. People shouldn't abandon the pursuit of becoming saints, who are the pioneers of conquering difficulties and frustrations. People shouldn't abandon the pursuit of becoming saints because of secular things because the filthier the society is, the more saints we need to offer us guides. The saints are responsible for the essence of human thoughts and the bright future of human race. Everyone should emulate the saints and try to be the saint of our age.

(2) Utilitarian knowledge. Utilitarian knowledge is widely accepted as a kind of practical knowledge. It has been widely emphasized and accepted since the time when it was created. The learning of utilitarian knowledge is through people's life, because people's lives basically rely on benefits, and human beings are living for benefits. Sima Qian said, "Hustling for benefit, all come; bustling for benefit, all leave."①

Human beings are basically learning and applying the utilitarian knowledge while they are young or even when they have become adults because they need the material basis for living and the accumulation of wealth, and also they need fame and fortune.

(3) Health knowledge. Longevity is the permanent pursuit of human beings. The degree of people's paying attention to health knowledge and its heritage is lying between ideal knowledge and utilitarian knowledge. People always don't have the sense of urgency in learning health knowledge and their emphasis is far less than what the bodies need.

Human beings do not pay much attention to health knowledge while they are young. When they are old, especially when they get ill, they begin to emphasize health knowledge with their health condition starting to decline.

Learning and practicing health knowledge is a life-long engineering for the whole human being.

1.23.3 The fate of learned persons

(1) Ideal knowledge. A lot of persons who created ideal knowledge were basically unlucky while they were alive or at least were considered unlucky by others (not by themselves), but their knowledge were extensively approved of after their death.

Confucius and Mencius were not thought highly of while they were alive.

---

① Quoted from *Records of the Grand Historian* written by Sima Qian, a famous historian and litterateur of the Western Han Dynasty.

Their personal developments, realizations of their ambitions and the degree of being respected could not be compared with the materialists at that time. For example, Confucius was not put in an important position while he was traveling through all the kingdoms, and Mencius was not as popular as Zou Yan① and Su Qin ② at that time.

Other examples like Karl Marx, Lei Feng and other saints and heroes in every nation are all creators of ideal knowledge and noble spirits. "They had no glory at their times and tended to be looked down as fools by those self-glorifying mediocre persons." "The real life is always snobbish and it makes the saints suffer too much." They left hardships to themselves while alive and shed light on the offspring. The person who creates spirit is noble and everlasting. The society needs spirit and it can not exist without spirit. Spirit is eternal and the person who creates spirit is also eternal.

(2) Utilitarian knowledge. The capable persons who are utilitarian and pragmatic are more popular and more likely to realize themselves. They are able to achieve success one way or another because they have methods to please the persons who are utilitarian and pragmatic. Zou Yan and Su Qin mentioned above are two such persons.

However, utility is short-lived while spirit is everlasting.

(3) Health knowledge. People who study and spread health knowledge can be accepted, approved and respected extensively. They can lead comfortable and peaceful lives and their fate is moderate. They can also be respected after their death.

1.23.4 Classifications of human beings according to the knowledge they own

Since the three kinds of knowledge in the world—ideal knowledge, utilitarian knowledge and health knowledge—are all useful, human beings should learn and apply them all. According to their attitudes and degrees of practice of the three kinds of knowledge, human beings can be divided into three categories: the noble persons, the utilitarian persons and the vulgarians.

(1) The noble persons. The noble persons can be further divided into

---

① Zou Yan is the representative of the Yin Yang School and the Five-element Theory. He lived in Qi State in the late Warring States Period and wrote *Zou Zi*.

② Su Qin (? —284 B.C.), courtesy name Ji Zi, is a famous political strategist, diplomatist and tactician.

three kinds: a. Saints. They have been pursuing their ideal knowledge and noble spirits all their lives without considering the cruel realities. They have been living for all the human beings. They have boldness of vision, just like what the great learned man Zhang Zai① in the Song Dynasty said, "Ordain conscience for Heaven and Earth; secure life and fortune for the people; continue lost teachings for past sages and establish peace for all future generations." b. Philosophers. They come to realize the truths when they are mid-aged and keep pursuing the noble spirit to enlighten others. c. Smart persons. They begin to pursue noble spirit at the end of their lives.

(2) The utilitarian persons. They have been pursuing material gains all their lives, considering ideal knowledge and noble spirit as something empty, high sounding and impracticable.

(3) The vulgarians. They care for nobody else except themselves. They are absorbed in their own comfort, not caring about others' suffering. They care only for their own health, not caring about others' living or death. What's worse, there are some persons like beasts who sacrifice others' lives to serve their own health. They have no luck to do with ideal knowledge and noble spirit.

Socrates ever said, "The basic purpose of every kind of knowledge is to solve the problem of 'How do people live?'" The living human being should be a person first, a noble person second and a saint third.

1.24 Gain the results and try gain the process

Everything is a process. A life is a process, teased by others both at the beginning and at the end, teasing others in the middle part. Life is a large process composed of a lot of small processes. Every process is for or can have a result. Every process is a part of our life. If we dislike a process, we will dislike one part in our life. If we have too many dislikes, we will have more unlucky things in our life. Therefore, we should emphasize our participation in life and experience the processes so as to make our life colorful. Results are important, but the processes are equally important. a. There would be no results if there were no processes. Therefore, we should design our process and

① Zhang Zai (1020—1077), courtesy name Zi Hou, a famous ideologist and educator of the Northern Song Dynasty, one of the founders of Neo-Confucianism.

put it into practice well. b. Process can give us experience, interests, enjoyments, and also hardships, so we should deal with the process seriously. c. Process can bring gains to us, which are different from results, hence we must value process.

Pay attention to and experience the process of life. What is the most important for life is the process, not the result. What is the result of human life? It is death. No one wants this result and everyone needs a happy life. Leading a life means to spend a period of time, which is a process. Someone has been leading a happy life, but he has not experienced it well and has been seeking happiness at sea. For this reason, we must carefully experience our life because every detail of our life is worth tasting and summarizing.

Life is a process of self-cognition. We must know ourselves bit by bit in our living process, knowing not only our shortcomings, but also our potentials and energies. The process will last to the end of our life. The results of the self-cognition will appear not only in the end of our life, but also have been there continuously in the process. Gains in the process are much greater than those in the end of our life.

Life is in process. The active factors are all working in the process, not in the end of the life, whether we are officials or soldiers, rich or poor. It will be a good process if we feel happy and comfortable in it. The perfect life should include having experienced all the joys and sorrows. The meaning of life lies in the process. If the life in the process is successful, we can say we have led a successful life.

Growing flowers is also a process. We can mould our temperament, cultivate our loving heart, adjust our life, develop our intelligence, and broaden our knowledge in the growing process. Blooming is a good result which can make us excited, but the flowers can not last for quite a long time. The blooming of the broad-leaved epiphyllum is only a fleeting show; hence we should gain the results, but also the process. We appreciate in the process and we gain in the process.

1.25 Learn to affirm ourselves

Never regret. Regretting is not beneficial because you have made your choice with efforts at that time. Owing to the influence of the outer environment and their own mental state at that time, people tend to regret what they have done. They regret because it is irretrievable, and if it is retrievable, they will not regret. However, we can warn ourselves and remind others using the improper things we have done, it is unnecessary to regret. We need to look forward.

Some people are always regretting what they have done and always have psychological burdens, which is quite unreasonable. Things that people have done can be divided into two kinds: things they've done consciously and things they've done unconsciously. One may have done something consciously because he thought it over according to both the subjective and objective circumstances at that time and thought it was the right measure that he should take at that time. If he can do it again, he will do it in the same way. Therefore, it is not necessary to regret. There is no reason to regret things that we have done unconsciously because they are not our choice. Even the saints may make mistakes. Regret will cause repenting and repenting will cause illness. It is not worthy.

Learn to affirm ourselves and do not deny our past easily. Denying the past means denying the whole life. Affirming ourselves can increase our self-confidence and denying ourselves will make us repeat the same old play. One should not look down upon himself. The upper class people should treat the lower class people as persons and the lower class people should treat themselves as persons.

In fact, the social evaluation on persons may not be completely right; it can be prejudice sometimes. The recorded evaluation or the evaluation in people's brains is never really true. Only the person himself knew what really happened to him. Even the person himself may not fully understand his own ego until his death. One may not know how energetic and how talented he is while living. However, it is enough as long as one feels he is holy. Affirming ourselves is never too excessive.

1.26 Look forward when there is a way and turn back when there is no way

Think about the future while young and recall the past while old. The young people are thinking about the future because they have only walked a short distance in life course, leaving little to recall, and they will have a long distance to walk. Thus they are only considering how to walk forward. The elderly have walked a long distance, leaving a lot to recall, and they will have a short distance to walk, so they are expecting little.

If an old person feels that he has a lot of things to do, he will have no time to recall the past. If an old person has nothing to do, he will do nothing but recall the past. Recalling the past is not the old persons' patent. Young people can also get immersed in recalling the past when they are confused or losing confidence in the future.

We should move forward bravely as long as there is a way ahead and we have to turn back if there is really no way ahead. For the people with strong believes, facing death does not mean it is the end of the road, but that they will continue their walking on the road that they have chosen, hence they march forward courageously and never retreat. There are alternative roads to anybody at any time, but the problem is that whether we like to walk on it or not. "Where the land ends is where the sea begins." Walk forward when there is a way and turn back when there is no way. Turning back can be another way and it is possibly a better way. "A good horse will never turn round to graze on an old pasture." This proverb means that a good man will never turn back to walk on the original road. This idea is individualistic heroism which is a manifestation of being obsessive, an expression without tolerance and the self-destroying act in human society. It is a silly action. Human beings will not predetermine a way of behavior and then they will never change it. They should be practical and realistic in order to seek truth from facts. It is sensible that once a mistake is made, we should correct it.

Chapter 2 Human Nature

Human nature is benefit-oriented and selfishness is the nature of human beings. Although human nature is the same, there are successes and failures among human beings. If you want to know the secrets in it, please read the followi

2.1 Don'ts

Don't cheat me; I will become smart.
Don't scorn me; I will succeed for it.
Don't attack me; I will be incited to realize my ambition.
Don't suppress me; I will advance rapidly.
Don't misunderstand me; I will become clever.
Don't regard me as enemy; I will revolt.
Don't ask me for things; I will become generous.
Don't give me misfortune; I will become persistent.
Don't hurt me secretly; I will learn self-protection.
Don't frame me up; I will show brilliance.

Don't give me ease; I will become decadent.
Don't take pity on me; I will be dependent.
Don't make me superior; I will stop forging ahead.
Don't make me self-willed; I will be arrogant.
Don't pamper me; I will become selfish.
Don't always praise me; I will become frail.
Don't give me big surprise; I will become sorrowful.
Don't give me wealth; I will become lazy.
Don't protect me; I will make big mistakes.

Don't put me in greenhouse; I will be unable to soar in the sky.

2.2 Poison and nutrition

Miseries are poison for the weak and nutrition for the strong; superiority is poison for the profligate and nutrition for the ambitious. Therefore, everything can be poison for the weak without ambition and nutrition for the strong with great will.

2.3 Human nature

People have many inherent natures, including imitation and comparison.

2.3.1 Imitation

People even imitate when they are in gene state, so antenatal training is very important. When people are born and especially when they understand the world, they show the nature of imitation obviously. As people grow up, they keep imitating. The objects of imitation are all kinds of people's and animals' knowledge, skills, games, and behaviors. In most time, people don't know whether the imitation is right or wrong. As is known to all, the wolf child who is different from social people imitated wolves because he grew up with wolves. Human knowledge can be divided into two parts: one is the imitation of ancients; the other is the development and innovation of modern people. With the accumulation of knowledge and civilization, people imitate more things so that the base of civilization is more profound. People keep developing in the process of imitation.

Because imitation is the nature of human beings, the objects imitated are especially important. A good behavior can be imitated by people, so can a bad behavior. It is glorious to create behaviors or things that should be imitated; it is harmful to create behaviors or things that should not be imitated.

2.3.2 Comparison

People are good at comparing things naturally and habitually, at any time anywheres. When we get up in the morning, we compare whether we sleep better or worse than yesterday; we compare whether the weather is better or worse than yesterday. When we meet people, we compare their faces, their dressings, their state of mind, their positions, their incomes, their wealth, their spouses, their children, etc. When we meet unfamiliar persons, we compare their appearances. When we go shopping, we compare the quality, the color, the style, the price of goods, etc. When we have dinner, we compare color, aroma and taste of the dishes. When we receive service, we compare the attitude of the service. When we watch a football game, we compare who plays better and who plays better or worse than before. Students compare scores and which teacher teaches better. In a word, we compare everything consciously or unconsciously.

We compare with each other, so people are divided into the superior and the inferior. The superior keep mental balance, and feel proud and happy. The inferior lose mental balance, and feel unfair and self-abased. Therefore, it is natural that most people want to do better than others. But this state of mind is selfish and uncivilized, which needs to be improved and overcome. There is no difference between people. We feel different just because we have not compared with each other completely.

2.4 Barbaric people and civilized people

Human behaviors can be divided into barbaric behaviors and civilized behaviors. People can be divided into barbaric persons and civilized persons.

Barbaric behaviors: Violate others first or hurt others on the excuse of others' faults. We shouldn't do these barbaric behaviors like beasts.

Barbaric persons: Once the barbaric behaviors become one's mainstream actions or even habits, he will become a barbaric person.

Civilized behaviors: Be friendly and cooperate with others. Never violate others.

Civilized persons: Obey civilized behaviors. Civilized persons can fight against others' violation.

If common people are barbaric, the harm is limited. If leaders are barbaric, the harm is huge. People who cause troubles and wage war are barbaric persons. We hope that there are fewer barbaric persons and people all over the world will punish barbaric persons.

2.5 Self-respect is human nature

Self-respect is human nature and necessity. Self-respect has two meanings. One is that we should respect instead of depreciating ourselves; the other is that we don't allow others to depreciate ourselves. Dignity is a formed respectable style.

Dignity is divine. People can suffer hunger and cold for dignity. Even selfish people want to pay for dignity. People would rather die for dignity. There is a sentence in *The Book of Rites*①: "A scholar prefers death to humiliation." The importance of dignity ranks top in human relations. Dignity is divine and supreme. Freedom and dignity are the supreme pursuit and the minimal requirements of humans.

Dignity is inherent. People need dignity as soon as they are born. Maintaining safety, benefits and fame is human nature. We should also maintain others' dignity consciously.

2.5.1 Dignity is universal

Everyone has dignity. As the saying goes, "Face is as important to the man as the bark is to the tree." Only the man you look down upon will abuse you. This shows the universality and importance of dignity. Everyone wants others' approval and social recognition, and needs face and dignity. Therefore, we focus on comments by the leaders, friends, and lovers, etc. These are all the manifestations of dignity and self-realization.

2.5.2 Special group's dignity can't be ignored

The dignity of the poor can't be underestimated. The poor have little wealth, but this does not mean that their dignity is devalued. The receivers' dignity should not be devalued, either. Giving doesn't mean having the right to humiliate others. The receivers need dignity. The dignity of those who are criticized shouldn't be violated. Criticizing others for the best doesn't mean you can ignore others' dignity. The children's dignity can't be ignored. Dignity

---

① *The Book of Rites* is a collection of texts describing the social forms, administration, and ceremonial rites of the Zhou Dynasty as they were understood in the Warring States and the early Han periods of China.

should be the first consideration when talking about children's education.

2.5.3 Maintaining people's dignity is the magic key of establishing good relationships

Self-respect is so important that if we want to establish good relationships, the wise practice is to maintain others' dignity consciously. Maintaining others' dignity is the first magic key of being in harmony with others. Maintaining others' dignity should become a habitual behavior. Say what we should say, and ask what we should ask. If we want to establish good relationships with others, we should make others know that they are very important.

2.5.4 Curve of dignity

The need of dignity grows as the improvement of living conditions. When living conditions are bad, the need of dignity is little. When living conditions develop, the need of dignity grows rapidly (see Figure 2.1). When we are well-off, the curve of dignity development is nearly parallel to the axis of dignity (Y-axis) and vertical to the axis of living conditions (X-axis). This shows dignity ranks definitely first when we are well-off. Dignity isn't influenced by improvement of living conditions and increasement of wealth. Therefore, in the well-off modern society, dignity ranks top.

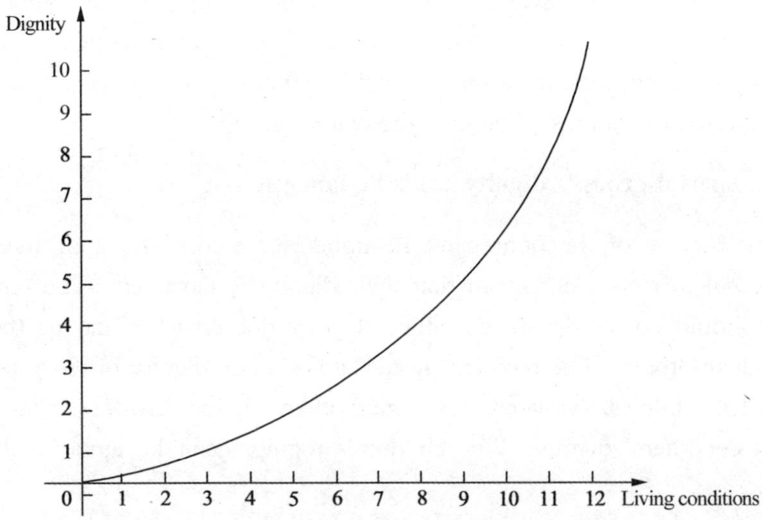

Figure 2.1 The curve of relationship between dignity and living conditions

Chapter 3 Ways of Life

Everyone is living in the society and it is the most difficult thing to conduct oneself in society. Human relationships are very complicated and there are many strategies in them. If you want to know them, please read the following.

3.1 Imperfection is absolute

Nothing is perfect in the world. As the saying goes, "Gold can't be pure and man can't be perfect." And it goes in another saying, "Men are subject to sudden changes of fortune as the moon is subject to the changes of being full or crescent, cloudy or clear."① The moon, which is round itself, always represents some changes to us. Should the moon or human beings be blamed because of those variations? No, nobody should be blamed, because it is the fact that all round things do not have to represent being round, and people do not always deem it as in a full round form unless it fully shows itself.

Man's body is imperfect. None of us has the organs in absolute symmetry form, and no one has a perfect organ.

Man's mood is imperfect. On one hand, the expression is imperfect, that is, people are unwilling to fully express themselves. On the other hand, although one has tried his best to be perfect, it is far from meeting the recipient's satisfaction. It is rare that both sides are perfect; therefore, imperfection is normal and absolute.

Imperfection is a kind of motivation. People have dreams because of their

---

① This poem, called *Prelude to Water Melody*, is from Su Shi, the famous poet in the Northern Song Dynasty. He created this poem in 1076 when he was in Mizhou, whose topic was missing his younger brother Su Zhe.

imperfections, and even one day when their dreams come true, there are still imperfections. Just like the one who is dissatisfied with his present life, he makes his pursuit of the goal, but after meeting his goal, he finds he will not be satisfied at all. "A man whose heart is not content is just like a snake which tries to swallow an elephant."

Handle any kind of relationships appropriately. As the saying goes, "Man is well connected to social relationships."① Therefore, everyone has many relationships to be well dealt with, such as, family relationships, neighborhood, colleague relationship, etc. Our life will be filled up with sunlight if we handle these relationships in an appropriate way. We need digital programs when we deal with natural science and obscure programs when we deal with social science. Natural science should be taken seriously and precisely and cannot be done casually with; as for social issues, especially contradictions among people, it means abstaining from it if we like to handle them directly with quantitative procedures. Obscure programs must be correctly used if you want to deal with social problems appropriately. It is also a kind of social science. There are some Chinese sayings, "Where ignorance is bliss, it's folly to be wise." "It is impossible to be good parents-in-law if you do not learn to play a fool sometimes." "Do big things with principles, do small things with special styles."② Life is not perfect itself, so why should we ask others to be perfect?

One must be disappointed if he measures feelings and emotions of others and his with a standard ruler; one will be away from true lovers and friends if he does so with love and friendship; one will have no friend if he demands others with a high standard. It is the most sensible way to treat everything with an attitude of imperfection, pleasure and forgiveness. The one who asks others to be perfect is not perfect himself.

---

① Quoted from *Capitalism* by Karl Marx, which is about the nature of human beings.

② To put it more specifically, for major issues, we should see the overall situations further, while for minor issues, we should leave some space for others to develop themselves.

3.2 Take precautions against barefaced compliment; make friends with the one praising behind your back

Lao Tse says, "Creditable words are not beautiful, and vice versa."① Confucius says, "He who gives fair words feeds you with an empty spoon."②

Occasionally, giving a barefaced compliment may result in flatter-compliment and sarcasm. It often makes the hearer feel uncomfortable because only encomiasts are interested in the laws of licking others' boots. Those honey-mouthed encomiasts to your face are dagger-hearted, and chicken-hearted surrenderors in the mind are flunkies.

However, praising behind one's back represents fairness, honesty and love, which makes people think it is true and the hearer has a great esteem for the speaker's character. One's words go fastest among other things, especially mean words.

A public judgment is a dime a dozen, while a private comment is worth much.

3.3 Defense is the first

3.3.1 Defense actively and keep it on

Considering our human nature, predicting and defensing the possible threats and hurts caused by what we've done and what we've met is the initial thing in our life. As the saying goes, "Harbor no ill intention against others, but never relax vigilance against evil-doers."③ Taking precautions against everything, because anything is possibly harmful to you. In other words, defense is the first, whether things we face are good or bad, live or dead. Otherwise,

---

① The sentence means that true words are not fair-sounding because they are not processed. In other words, we can say that words or expressions that are full of rhetorical speeches may be not true.

② Quoted from *The Analects of Confucius*. It means that a person always with sweet words is not with a benevolent heart.

③ Quoted from *Discourses on the Roots of Vegetables* whose author is Hong Yingming. It teaches us that we should not have the victimized heart and do not readily believe others.

dangers are coming soon. In Chinese poem "Jingxi River"[1] written by a poet named Du Xunhe in the Tang Dynasty, he said, "The Jingxi River is well-known for its dangerous rocks in the water, but there were no persons drowned all the year because they were very careful while crossing it; however, we often hear that there were persons drowned at some plain water area where there are no dangers at all."[2] Humans are fragile but in constant motion and constant change, so is our world. Therefore, it's hard to avoid accidents or hurts. If we meet with danger, that indicates danger comes to the surface, while if we don't, that indicates danger hides itself away. Dangers are everywhere, so we should take precautions anytime anywhere. Only when our safety and profits (interests) are guaranteed not to be threatened can we gain progress and success. Here are some examples:

Work. You may get hurt in office, let alone those outdoor jobs, construction works, mineral mining and geological exploring, etc.

Social communications. Social communications have various forms and a wide range, such as, excursion, long trips, parties, etc. There are possible dangers in anyone of them, and even too much alcohol makes one dead in the wedding banquet.

Doing business. Interests go along with risks. Get rid of risks and then gain profits.

Fighting. Don't take winning or losing as a trifling matter. We should be observant to avoid failure rather than prepare a triumphal feast so quickly.

Please always take precautions against a calamity. People usually consider things seriously at the beginning, but gradually they slack off and overlook potential dangers, then there will be problems. If one can always keep guard system in his mind, he will reach the goal of checking erroneous ideas at the outset.

3.3.2 Take others as a mirror and nip an evil in the bud

Bad things which happen to one person today may happen to you tomorrow. There is a saying going like this, "A life cut short one moment makes

---

① "Jingxi River" is a philosophical poem written by Du Xunhe, a poet in the Tang Dynasty.

② The true value of "Jingxi River" is to warn people that we should be prepared for danger in times of peace.

one sigh, who would have known it's her turn next to die?"[①] We humans are alike, and our lifestyles and ways of thinking are basically the same. There must be reasons why other people make a mistake or have a bad fortune. You will be the next one unless you learn from others and take precautions against calamity. Hugo says, "Caution has much more power than audacity."

Of course, we should attach our gain or loss and personal safety to the social interests, giving top priority to people's interests and sacrificing ourselves for our country if necessary. This is a sacrifice despite that we have known the dangers involved, so it is not meaningless.

3.4 Communication is a magic key for closer interpersonal relationships

Fish live in the water, while human beings live in a social environment. The society is made up of individuals and their interpersonal relationships. Individuals are variable and changeable. Understanding each other is essential for a closer relationship while communication is the only way of achieving this goal. Based on mutual understanding and mutual respect, communication is the act of transmitting the information to others which they should be informed of in a comprehensive and effective way.

Good communication between family members will contribute to a harmonious and happy family; good communication between colleagues will result in a win-win cooperation; good communication between merchant and client is of great benefit to a successful business.

There is a story going like this.

> Due to business need, a man hurried to a copy shop to copy a color pattern. The pattern copies were not good enough when the traditional method was used, so the operator tried various kinds of other methods again and again and finally got the desired results. What impressed the man most is that the operator always told him what he had done, what he had not done and what method he would try next at every step. Every time this experience occurred to the man, he was satisfied with the operator's patience and confidence and wanted to drum up more business for him. However, if the operator hadn't had a good communication with his customer during

---

① Quoted from *A Dream of Red Mansions* which contains pessimistic mentality, nihilism attitude and cynical realism.

copy operation or hadn't let his customer understand every step and reassured him, things would have been different though the result was good enough.

3.5 How does a good mood come from?

"Wish you a good mood every day" is a popular wish. But how does a good mood come from? Here are some principles.

(1) It is necessary to be polite to others and keep in reciprocity with others. As an old saying goes, "One would not be blamed for being extra-polite." It is a fact that if you are extraordinarily respectful to someone, he will be more respectful to you. Generally speaking, being respectful to others is much better than being cold.

(2) Live in peace and try to avoid conflicts. Peace and harmony are the bases for a good mood, while conflict and quarrel often result in regret.

(3) Talk face to face with somebody and criticize others in a proper way. Do not comment on others' errors while chatting. No one would like to transmit face-to-face praising words but bad comments on others behind them run faster than good ones. And it would be unhappy if the bad remarks reached to the ears of people involved. It is not appropriate to criticize others with harsh words. We should criticize others at proper time, in a proper place and in a proper way, putting others' acceptance and understanding at the chief place. Playing jokes should not go too far, otherwise, it'll hurt others.

(4) Let nature takes its course and make the best use of the circumstances when necessary. Be optimistic about how things are going on, and be easy-going while speaking with others. Persuade others in a harmonious way.

(5) Pay attention to the interests of the whole and do not complain. Complaining is no use, so just stop it.

(6) Rise above jealousy and eliminate the hatred emotion. Do not only learn from people we don't know, and do not envy people around us. We human beings are somewhat jealous, and we are jealous of the familiar persons, but look up to the unfamiliar one far away from us. Jealousy and resentment are just like fire, while good mood is just like water, so they are not compatible with each other. Therefore, if you want to have a good mood, you should try your best to avoid jealousy and hatred. There is a Chinese proverb that "Forgiveness is the release of oneself and also the caress of others." According

to a broad scale research at American Duke University, intense hostility will increase 29% the risk of patients dying of heart attack, and 50% more among people who are 60 years old or under 60.

(7) Be good to others and always cultivate your mind. You may feel relieved when you help someone out, meanwhile, you will receive gratitude and reward in return. People should cultivate their minds actively to avoid disturbance from this mundane world.

(8) Feel deeply grateful to others and keep returns in mind. One should have the heart of gratitude. Always getting nurtured and pampered from others in life, one must be grateful and have an indebted heart forever.

(9) Be tolerant to others and bear with oneself. The one who have big mind and compassion will be greatly blessed. With forgiveness but no hostility, one will not only get relaxed or have a good health, but also get a good interrelationship. If we find more kindness from others, we will have more friends; if we find more faults from others, we will have only hostility. Do not be so harsh to yourself, forgive the occasional mistake, and avoid self-accusation for a long time.

(10) Adjust yourself, and defuse the trouble. Life is full of troubles. To lead a happy life, one should adjust himself by learning how to face favors and humiliations with composure and not think too much on gains and losses. Free oneself from worldly problems, then one will naturally feel good. According to the Balance Theory, one needs to adjust himself due to the imbalance inside his heart.

Try to look for happiness, and try to find the spiritual sustenance. If career is your biggest pursuit, you should focus on your career, seeking the easiest, the most economic and the best way of achieving the great goal; if beauty is your biggest pursuit, you should do in the same way, focus on it, remove the unpleasant thoughts and make yourself happy.

Wc should jump out the circle to regulate our emotions. With numerous ideological circles, it can be said that how many things you have experienced and cognized, how many circles there will be. When you are tired or bored in one circle, just jump out and go to another one where you may feel cheerful and happy to release your tiredness and vexation by giving mind a free journey.

In short, there are three words: goodness, kindness and lenience. Goodness is being kindhearted; kindness is being peaceful, compatible and harmonious; lenience is being lenient to others. If you were a person with characteristics of goodness, kindness and lenience, you would not have vexation at all.

When goodness, kindness and lenience come together, it forms one word "love". Universal love is to love all the people sincerely and naturally. Love creates harmony, infection and power; love makes self-consolation, self-pleasure and calmness; love produces happiness and health.

It is not easy to love truly, purely and consummately. Love needs foundation, open minds, willpower and perseverance. Some people love their children in the way of beating and scolding, some with pampering; some people love students in the way of punishment and sarcasm; some people love others in a way without considering respecting others. These are not fully pure or perfect love.

Moods come from interacts between people and the society, and the style of handling matters determines the mood.

3.6 The highest level of competition

The highest level of competition is to compete for harmony. The winner is the one who can be the most harmonious. Its significance is reflected in the following aspects: First, one can save energy, manpower, resource and money cost in fighting against other people; second, the negative loss of confrontation and friction with the opponents is eliminated; third, people will gain more partners, more opportunities and larger areas of mutual cooperation. As a proverb says, "Business has no difference and harmony makes alive." Therefore, whether we can win the competition depends on how much we have spent on "harmony". The winner of competition must be the winner of harmony.

3.7 Large-scale production of happiness

Happiness is what all people need, so many greetings wishing happiness are used, such as "Wish you happy!" "Wish you a good day!" or "Wish you joyful forever"...

All people need happiness but always lack it. Can we produce more happiness? The answer must be "yes".

3.7.1 The elimination of vexation

(1) Sources of vexation. Vexation results from being away from ego. The further we are away from ego, the more vexation we will have, and vice versa. Ego has two meanings. One is leading oneself: to do whatever one wants to; the other is self-living: to stop being worn out with pandering to others. Implicated by environment, interfered by others, lacking of autonomy and freedom are the causes of melancholia. Of course, too much self-centeredness is unacceptable in society. Those who chase for various benefits or even with growing greed will stay far away from ego, and then vexation comes along. Here is a parable story:

> A group of kids were playing in front of an old man's house. A few days later, the old man couldn't stand any more.
>
> He came out and gave every kid 25 cents and said to them, "You guys make here very lively and I feel much younger, here is my thanks."
>
> The kids were very happy and came again the next day, playing as before. Then the old man came out and gave every kid 15 cents this time and explained that he had no income at all and could only give them so little. These kids left still happily.
>
> On the third day, the old man gave every kid only 5 cents. The kids got angry this time, "Oh, only 5 cents, you couldn't know how hard we are playing!" They swore that they would no longer play for him.

Motivation can be divided into two kinds: internal motive power and external motive power. When internal motive power dominates, we are quite clear who we are. When we are driven by external motive power, we will be swayed and confused, then become its slaves.

In the parable, the old man's strategy is very simple. He transformed the kids' internal motive power—playing for their own pleasure into external motive power—playing for cents. While he manipulated the external motive power, he controlled the kids' behavior. Originally, the kids were only pleased to play there, but finally it became a burden and unhappiness. So happiness relies on ego, while vexation results from burdens.

Material determines spirit, but material world can't always meet the requirements of spiritual world, which is bound to produce vexation. If one must have good mood, he must eliminate the cause of vexation which is one's selfishness. Selfishness is never satisfied, and vexation is always accompanied by selfishness. As the saying goes, "A man who is contented will be happy."

(2) Harms of vexation. Vexation can cause illness. Our body will produce something beneficial or harmful when affected by internal or external motive powers. And vexation is sure to produce harmful substances to our health which we should release as soon as possible, otherwise, they will gradually accumulate and make us sick as time passes. With pleasant mood, physical mechanism can operate normally and body itself can regulate discomfort, at the same time, prevent and dissolve diseases. Therefore, good mood is the best doctor. "Without good mood, medicine is useless." Hardworking cannot hurt heart, while vexation will make people old. It's beneficial to think more about others, but harmful to worry about ourselves. Annoyances destroy our career. Because of them, we can't handle problems with calmness and screw them up. Annoyances disturb life. They make life disordered and affect life quality. A life filled up with annoyances is meaningless.

(3) Trouble-resolving. Vexation comes from senses and has effect on one's heart. How can we eliminate those annoying things? If we haven't solved them yet, here we have two methods: one is thinking, and the other is releasing.

1) Thinking: To completely eliminate vexation from heart

First, we should know that life is a process. If we can live to 80 years old, it is just about 20,000 days. Life is nothing but simple. Everyone should feel happy because we are now experiencing life. Even if we cannot get along well with other people, but it is distinctive, for we have no chance to experience it more. It is just like what Wang Fei, my daughter says, "You are happy because you are alive." During our life journey, we should continually sum up our own experiences and lessons, and analyze the unhappiness thoroughly, then find out the better idea to direct our later action. In this way, not only can we avoid annoyance but gain more reward and pleasure.

Second, we should know that everyone has his own vexation since we humans are alike. As the proverb says, "Everyone has his own problems and every family has a skeleton in the cupboard." In this world, we can't do whatever we want to and we can't have whatever we need. People cannot ask for what comply with our requirements, and it's senseless that we are not reconciled once we fail, because there is no perfection. Imperfection is reasonable, normal and perpetual. Too many troubles occasionally occur in our life, and some may not be your fault, but you have to take bad results or unluckily get hurt. These are unavoidable. So we don't have to care so much. Those who can't bare misunderstandings and slanders are losers. We should think about brighter and better things. Life is full of troubles, so do not complain but face up to them. The bittersweetness of life is just like the full moon once in a month. We will feel better if we look at it from another point of view. There is a story in *Grimm's Fairy Tales*.

> An old lady was very sad because her son died, even her soul was out of balance. She went to ask the Father, "Why did my son die? Can you bring him back to life?" Father said, "Well then, you may take a bowl and ask for some grains of rice from someone who has never suffered a death in the family. When you get ten grains of rice, your son would come back to life." Hearing this, the old lady got very happy and began to beg, but on her way from door to door, she was surprised that there was no family which had never lost relatives. Finally, she got nothing. After this, she realized that it was unavoidable that relatives would pass away for a family. By comparison, she felt balanced and then lived a happy life.

Third, we should know that life is made up of numerous kinds of things. Things are like water and instantly become the past. Life is a stack of things. After one is solved, another one appears. We don't have to complain or get annoyed with them. Don't worry and take it easy. If there is nothing to be bothered about, there is no meaning to live. If you are annoyed with things, that means you are annoyed with life. However, the real life is filled with good things and bad things; we are constantly solving one problem after another.

Fourth, we should know that we cannot justify whether the valuation of us is fair or not by ourselves. Whether others are fair to you or not depends on both sides. Whether the society is fair to you or not depends on the social environment. To judge whether others are fair to you in a reasonable way, you should firstly compare their background and motivation with those of what you have done and analyze those reasons, then you two need good conversations. If we could not do something well or were misunderstood as not good enough, but we were asking others to treat us well, otherwise we would get angry, then we were too selfish and self-centered. It is just like the injustice of the magistrates being allowed to burn down houses while the common people being forbidden to use fire even to light lamps. Compared with the ulterior or unexpected things we have thought and done, the unfair things we met are trivial. Shouldn't we feel lucky? Whether we are treated fairly or not relies on the proportion between what you have contributed and what you have got. Whether the society is fair or not will be influenced by the social environment. It is a problem to all humans not to us only.

Fifth, we should know that we cannot judge a hero on the basis of success or failure. Success is normal, so is failure. Whether one succeed or not is the combined result of self reaction and external reaction. Everything is inevitable, and failure is often not one's fault. It just should be like that. Don't be that greedy, and be moderate in fame and fortune, for few desires cause little trouble. Don't insist on something for everything has its own way to survive. Just take advantage of opportunities. A plain life should not be disturbed by extravagant demands. Wang Fu① said, "Good rank cannot make me respectable, and a humble origin cannot make me really humble... Gentlemen may not have much wealth, and little men may not be poor and lowly."② The book *Discourses on the Roots of Vegetables*③ said it well, "To remain indifferent, leisurely watch the flowers bloom and fade; whether to go or stay, calmly see the clouds roll and extend." Besides, Tseng Kuo-fan④ once said, "Take everything in the world easy and your mood will be stable. Heart cultivation needs few desires." Live a life sensibly, and keep the state of being detached and unrestrained as a canon. Compared with those great men, do we have as much sufferings as Sima Qian⑤? But he rose in despair, writing the famous book *Records of the Grand Historian*⑥. Are we aggrieved as Yue Fei⑦? He died a martyr's death. Do we suffer a heavy loss like Cao Cao⑧ in the "Battle of Red Cliff"? He just had a good laugh and cheered up again. Do we have

---

① Wang Fu, a political commentator, litterateur and progressive thinker in the Eastern Han Dynasty, was born in 85 and died in 163.

② If you are in a high status, you should take it as a kind of rank and honor; while if you are in a lower status, you need not be humble.

③ *Discourses on the Roots of Vegetables* is a book on quotations which discuss the ideas about accomplishment, value of life, ways of life, and escaping from the world. It is regarded as a famous book full of ethical thoughts of life for Chinese nations.

④ Tseng Kuo-fan, a famous politician, strategist, moralist and litterateur of contemporary era of China, one of the famous officials in the late Qing Dynasty.

⑤ Sima Qian was born in 145 B. C. and died in 90 B. C. He is a great historian, litterateur and thinker in the Western Han Dynasty, and he is also well-known as "the father of Chinese history".

⑥ *Records of the Grand Historian* is a historical book with the feature of history presented in a series of biographies and it is the most famous historical book in the history of China.

⑦ Yue Fei, who was born in 1103 and died in 1142, is a famous militarist, strategist and national hero of that period.

⑧ Cao Cao, who was born in 155 and died in 220, is an outstanding politician, militarist, litterateur and calligrapher in the later Han Dynasty.

three ups and downs like Deng Xiaoping[①]? He pointed the way of China's development. Do we endure hardships like the Chinese Red Army's Long March? They led China to the bright direction.

2) Releasing: To pour out vexation from our inner heart

Vexation is the vicious power forced by the external world. We need to find ways to eliminate troubles once finding them. Specifically, we have the following methods.

First, release yourself. We can release worries through mouth or hands, such as pouring out your bitterness to friends, writing them down, singing, reading and shouting loudly to the mountain or sea.

Second, adjust yourself. There are many ways to adjust yourself. Choose the way you like, such as having a rest or sleeping. After relaxing, you will be in good mood as your energy recovers and then annoyance will be reduced. Treating yourself to shopping or snacks, listening to music, watching TV or movies and dancing etc. can all be the methods of adjustment. When you are happy, you may comb, massage your face, ears, hands and feet, then keep them your habits. In later days, these activities may help you to bring back good mood. When you are unhappy, analyze causes of vexation and dig out the law to direct your later action so that you will be distracted and release pressure. Besides, we can go to the nature to release pressure, for example, climbing, swimming, and running, etc. Here is a story.

> There is a man in ancient Tibet. Every time he got angry he would run around his house and lands three times. Later, as he grew older, his house became bigger and his lands became more, he still ran as before. Once, his grandson asked with puzzlement, "Is there any secret in your running once you got angry, grandpa?" "When I was young, as I got angry or lost balance in heart, I would run. While running, I was thinking that since my house was so small and my land was so little, I had no time and energy to be angry with others. Once I thought about this, I would calm down and have great courage to work hard." said the old man. Then the little grandson asked, "But now you get old and rich, why are you still doing that as before?" "Because now as I am running, I am thinking that since I have a large house and large land, why do I have to care about those messes?"

---

① Mr. Deng Xiaoping, who was born in 1904 and died in 1997, is the leader of the second generation of the Communist Party of China. He is the great Marxist, proletarian revolutionist, politician, militarist, diplomat, the member of Communist Party and the Chinese People's Liberation Army, and the creator of Deng Xiaoping Theory.

Third, transfer yourself. Change your perspective, transfer your worries, resolve your distress and let it be.

Change your perspective. If you see the same thing from different views, you will have different mood. You will be in good mood if seeing it positively; otherwise, you will be depressed.

For example, there is half a pitcher of water. To the optimistic there is still a half, while to the pessimistic there is only a half left. Here is a story.

> An old lady was always depressed either in sunny days or in rainy days. The reason was that she had two daughters, one selling umbrellas, the other selling ice-cream. When it was sunny, she worried about her elder daughter' business; when it was rainy, she was anxious about her younger daughter's business. One day, a wise man told her that there was no need to worry like this every day. He advised the old lady to be pleased for her younger daughter in sunny days because the ice-cream must be on good sale and to be pleased for her elder daughter in rainy days because the umbrella must be on good sale. So she could transfer sadness into happiness. With this inspiration, the old lady got very happy. "How couldn't I think of that?" the old lady smiled.

Transfer your worries. Having thought too much upset things, one should try to divert the attention. If one has unpleasant feelings here, he may leave here for a second and put his energy into what he is interested in, then he will forget that unhappiness.

Resolve your distress. Everyone has his own distress and the point is whether he can lay it down or not, and if he can't, he will be crushed by it.

Let it be. Everything goes in its own way and can't be demanded. It is sensible to let it be and make the best use of it.

3.7.2 Create pleasure by ourselves

There are only three kinds of products produced in the world: human beings, materials and spirits. Pleasure belongs to spirit. People have been pursuing happiness and raising happiness index in a rapid pace today. Therefore, pleasure production should be balanced with material production. Happiness is surely more than pleasure. Pleasure is a short-lived feeling while happiness is long-lasting. Pleasure divided by desire equals happiness. The lasting pleasure is happiness.

Human is a combination of body and thoughts. Body is nourished by materials while thoughts are supported by spirit. Physiology determines body and spirit determines thoughts. Pleasure depends on the state of physiology and spirit. To make one happy, both his physical demands and spiritual demands need to be considered. First, we should adjust the physiology on the basis of material wealth; second, we should straighten out thoughts on the platform of spirit. As the proverb says, "Clever men contribute and wise men desire, while the impotent needs nothing." To the wise people, spiritual life should be focused on, while to the others, the material life is more important.

(1) Strive for happiness from the objective world. First, we should grasp the chance and work hard to gain more achievements. Second, we should do whatever we want to, which can help us satisfy and relax ourselves and create more opportunities of happiness.

(2) Strive for happiness from the subjective world. Produce our own theoretic foundation of happiness. The world is material and spirit is subordinate to material. Material determines spirit while spirit reflects material. Material world is the first world while spiritual world is the second. Natural sciences are more dependent on the first world and social sciences are more dependent on the second. Everyone has these two worlds (they can also be called objective world and subjective world). The objective things cannot change as the demands of subjective world. Only in the second world can you find ways to have long-lasting good mood or happiness. From the view of philosophy, everything in the second world is decided by the first world, however, as a world, the subjective world has its own independence because the essence of spirit is material. There is nothing else but materials in the universe. Since spirit is material, spirit can decide itself and spirit itself can produce spirit. Happiness and state of mind belong to the second world; therefore, they can exist independently and not be influenced by the first world. We can conclude that things like good mood and happiness are completely controlled by our human beings, no matter when you want and how much you want. That is to say, if you think you are happy, then you are; if you think you are not, then you are not. If you want to have good mood, then you will have. "Sadness, worries and pains are all caused by ourselves." said Rousseau. However, it is a process that materials produce spirits or spirits create spirits. They cannot be produced without reasons, but need knowledge, understanding, study, practice and experiences.

Strive for happiness from subjectivity. Vexation and unhappiness always result from the feeling of incapability, therefore, we need to find our own advantages to improve ourselves and balance the mind. Keep good habits to list the merits or achievements and analyze them in daily life. Only in this way can we enhance our confidence and enjoyment. Human beings are spiritual animals and live a spiritual life. Spirits belong to us, and also can be controlled by ourselves, so lead a life what you want it to be.

Good mood is in our control. We should not let our mood be controlled by others. We are happy if someone is nice to us, while we are sad if he is not. Why are we under others' control? We are living for others' living. It is a sign of self-doubt, dependence and weakness. Why are we manipulated by others? Why shouldn't we be more rational and less emotional? Why aren't we our own masters?

Man struggles upwards while water flows downwards. So being bitter first and sweet later is easy to be accepted, otherwise, it will be difficult to be accepted. Therefore, we should do something difficult first, eat something unpalatable first, suffer first and solve unpleasant things first. Don't leave unpleasant things which are the barriers of happiness to the last minute, for if they are all solved at first, the left is completely happiness, otherwise, only worries and sorrows are left. Being bitter first and sweet later is the living style of immortals, while sweet first and bitter later will be the cause of agony.

Elongate happy chains. Celebrating, mentioning and thinking a good thing for several times are effective ways to multiply and produce happiness.

Live a life with no excuse. "A good man is seldom uneasy, while an ill one never be easy."① Be optimistic towards everything which is helpful to success. For example, arriving at the traffic lights when they just turn red, the pessimistic says, "Oh, unlucky!" The optimistic says, "Well, I am No. 1." Let the heart be a little wider, and illnesses will be farther away. Have a stronger ambition, and more progress will be made. Being narrow-minded may cause vexation. Feeling old in the heart for a while results in long-lasting oldness in body. Try to be more humorous, more appealing, easier, and more modest. Control your own mood, cheer up and take actions immediately. A good beginning is half done.

(3) Create a happy atmosphere. Good mood is an important part of im-

---

① Quoted from *The Analects of Confucius*, which is familiar to common people from ancient times to the present. It means that gentlemen are broad-minded while vile persons are haggling over every ounce and worry about their gains and losses.

proving the quality of life and constructing a harmonious society. So we need to create a happy atmosphere in society.

Create a happy working environment. Unit leaders and colleagues should pay attention to setting up a good working environment. It is not friendly of those who do not bring happiness and good mood to others. A cheerful working atmosphere which unit leaders created is approximately equal to spiritual welfare for workers.

Create a happy atmosphere when you get along with others. Try to be a man who can always bring happiness to others whether when you're first meeting each other, working together or separating from each other. We should respect other people's hard work, advantages and personality; meanwhile, learn to praise others because people all like to hear admiring words which can help enhance their important role. Inevitably, a leader likes to promote persons who admire him as employers. People easily accept compliments, so they like to tell their colorful experiences—something like the story of forcing five passes and slay six captains. Everyone likes humors and jokes which make them relaxed and happy. Eric Bressler of McMaster University in Canada has studied that the importance of humor is different for men and women. Woman is attracted to a man who often makes her laugh; man prefers a woman who laughs for his humor or jokes.

Seeking common points while reserving difference can make interpersonal relationship comfortable. People have different words and deeds because of their different background, knowledge training, character, temperament and way of thinking, which is an objective law and cannot be violated. Therefore, the method of seeking common points while reserving difference is the only way to keep good interpersonal relationships.

Help others produce happiness. Happiness, which is mutually inflamed, is produced with others' cooperation, especially the ones who company you all the time. First of all, one should be happy himself, which will affect people around to be happy. Second, one should be kind to create a happy environment, produce happiness for others and help others produce happiness. I produce happiness for others, they produce happiness for me, and we help each other produce happiness. While producing happiness for others, we are very happy ourselves. Happiness is a double-edged sword: good to the kind while sad to the wicked, because the wicked do not want to see other people happy. A person is painful himself, but may smile to others; a person cannot do physical work, but he can produce happiness.

Kindness is happiness. The highest pursuit for human nature is kindness. Religions advise people to do good deeds; the dominant idea of society is to guide people to do good deeds; government advocates people to do good deeds, and the entire human race are pursuing goodness, advising goodness and doing good deeds. Doing good deeds is the great pursuit and also the lowest requirement for human beings. Being kind and compassionate to others will make people have pleasant feelings, cheerful moods, positive attitudes and enable people to live longer.

Doing good deeds resonates with human nature. On one hand, doing good deeds will let people have the sense of superiority complex, which is beneficial to improve self-confidence; on the other hand, doing good deeds, with no psychological burden, will be highly honored and good for physical and mental health. We should be good to others' words and deeds. Especially when other people's behavior surprises us, we should firstly accept it, then think about the truth. Doing good things for the benefit of others is also doing good deeds for us and taking care of others is also taking care of ourselves. "Kindness has become a spiritual need."

Misdeeds distort human nature. On one hand, misdeeds cause sense of guilt and some heavy ideological baggage; on the other hand, mind would be disturbed and tend to be abnormal, which would result in psychological changes or disease due to the long-time anxiety. "He who is unjust is doomed to destruction."

(4) Cultivate character and develop a good mood. Excavate the beauty around actively. There are a few times a day of widening the eyes to appreciate everything and to accept them with a broad, expansive mind. For example, grass, leaves, buildings, people, and anything you meet are beneficial to you. All human behaviors, including thinking, will bring substances good or bad to the body. These substances would impact one's physical and psychological behaviors. So keeping a habit of pleasing yourself every day will gradually develop a positive and happy attitude. Bacon says, "There are morning glow, dew and flowers blooming year round in the hearts of those who appreciate beauty, but only icy land, dried sea and barren hills to the indifferent."

Upgrade your ambition and taste. Having higher ambition and taste will naturally let you ignore some trifles, but put more time and energy in the pursuit of the noble cause.

Cultivate various interests. Life should be colorful itself, and dullness will certainly cause illness. For example, developing a regular, habitual singing habit can greatly regulate and improve your mood. Besides, running, swimming, dancing, playing chess, playing cards, reading, writing, painting, calligraphy and playing the piano, etc. are all good ways to cultivate your character.

Be amazing and awesome. Other people will be happy for some happy things, so will we; other people will laugh when they hear jokes, so will we. But what is uncommon for other people is that we can still be happy having no happy things at all, which is extraordinary.

Producing happiness should start from children when they are young. We should cultivate their good mood and happiness carefully and set up different training environments because good character is formed in childhood.

A good mood is a nurse, a doctor, a regulator, a happy sensor, the wealth, and also the lubricant of a harmonious society. It is not money but better than money. It is not health but it can bring health. Marx says, "Having a good mood is better than taking good medicine to relieve pain and sufferings." Let's take actions immediately to produce more happiness.

3.8 On friends

What is friend? Friends are persons who are second closest to you after relatives in social life. Friends can be divided into three kinds: a life-time friend (a Damon-and-Pythias friend), a season friend and a reason friend.

Life-time friends, who have noble characters, as good as gold, are persons your life can be entrusted to. In the 4th century B. C., an Italian man named Pythias was going to be executed. He wanted to see his faraway mother before he died. Moved by his filial piety, the king agreed with him on condition that he must find a person to guarantee that he would be back. Finally a friend of Pythias named Damon agreed to be his guarantor. While by execution date, still no message about Pythias arrived. However, when Damon was brought to the scene by the hangman, Pythias came galloping in the rain, he shouted, "It's me, I'm coming!" Here, Damon is Pythias's life-time friend.

Friends in a season, which are formed by benefits, form the majority of friendships in nowadays environment of market-oriented economy. Social relations are the relationship between benefits and emotion, and are often built on the basis of benefits; therefore, friendship is a kind of relationship about benefits, yet a more intimate one, along with conspiracies and struggling for fame and fortune. Benefits are often more important than friendships. As the saying goes, "We have no eternal allies, and we have no perpetual enemies. Our interests are eternal and perpetual." The intimation between friends can be measured through dollars and cents. Friends turn against each other because of one hundred dollars, which means the friendship is worth only one hundred dollars; friends turn against each other because of one thousand dollars, which means the friendship is only worth one thousand dollars. To our surprise, the amount of money can test a person's conscience. Friends in a season are just the ones who can be measured with dollars and cents.

Friends in a reason are the persons who may want to harm you from the beginning, or perhaps are extreme individualists. In short, they are dangerous for you. When you meet an unclear condition, they may betray you after asking for your help. This kind of friends is just like your own shadow. He might follow you when you walk in the light; once you go into the dark, he would immediately leave away.

Friends are often the most dangerous persons to you. The persons who are closest to us are sometimes the most dangerous, too. There are three reasons. First, mutual infection may be caused because of the biological chain. Second, friendship may develop in the direction of rupture for the emotional collision from time to time, and friends who have broken their bonds of affection are similar to enemies. Third, persons who know you best also know how to hurt you most. So, some people say that friends are a ticking time bomb buried by your side, are the ones who are jealous of you, or even hate you, who rely on you if you are profitable and stay far away if you are unprofitable on stay in the way, and who desert you when you are in sore trouble and distress. Despite that these words are a little extreme and cannot be generalized, they ring an alarm. As an ancient saying goes, "Life with a bosom friend is sufficient." That means it is hard to find a bosom friend and few people can get one. So be cautious to choose one as a friend, otherwise, you are on the way to the jaws of death. "Play chess with the chess master and communicate with celebrities."

Trifling things, like rulers, help us to know people, while big events, like mirrors, help us to distinguish friends. "A friend in need is a friend indeed." Major issues of principle are the touchstones of friendship, but there are not so many big issues during one's life. Recuperation, reading, learning, and ordinary time must also be accounted for the bulk of the living hours even in the magnificent life, so it's really difficult to identify a person's reliability. However, the way a person deals with trifling things will influence the way with big events; the way he treats others will influence the way he treats the family members. Nothing is trivial in the world.

"The friendship between men of virtue is pure like water, yet affectionate; the friendship between men without virtue is sweet like honey, yet easily broken." "Those who do not consciously get close to you are most likely your true friends." "True friendship can not be interfered with personal gain, and only those who uphold such a principle will be possible to have real friends." "The way to make friends is to know him first, then gradually get closer." Perhaps these words could give us a lesson about making friends and knowing them.

The number of friends is constantly increasing in human communication. "One more friend, One more way." "More friends, smoother path." But you may encounter some bad friends from time to time, who should be promptly cut off in order to ensure your own safety and clean environment to live.

3.9 Equality brings with harmony, and difference causes contradiction

People are never equal, and the idea of equality is just a dream. For example, countrymen and city people, subordinates and superiors, poor and rich, mental workers and manual workers, etc. are so different that there is no possibility for them to be equal. Is there any equality in the history of mankind? Equality is just the demand of human nature, the wish of idealists, and the goal that people always aspire to.

Equality is helpful to relationship construction. Being equal on everything is impossible, but being equal at a certain time and in a certain place can be realized. It is an active behavior that people want to make friends with able persons or celebrities for the sake of interests. In fact, people instinctively tend to get along well with those people who are not perfect enough. Because there is kind of superiority, and it's easy to balance the mind and weaken the resistance of acceptance, or because everyone has shortcomings, and it is easy to get close with normal people. Equality brings with harmony. As the saying goes, "Water which is too clean has few fish; he who is too critical has few friends." Extra smartness and perfectness can be a threat. It will make contrast with other people; hence they would stay away from you.

Being equal with each other is easy to get close and cooperate with others. For example:

(1) Tell your own suffering childhood, which can cause other people's compassion who has the same experience.

(2) It is easy to succeed if your dress is similar to those who are going to negotiate with you. For example, it is better to choose a working uniform than a suit if you talk to the foreman in the construction site and your only choice in a formal occasion is a formal dress.

(3) Similar views, likes and dislikes can make a closer relationship.

(4) Good contacts with others is helpful to find common ground, hence it can increase degree of intimacy.

(5) Pay attention to the place, the time and the degree when you are communicating with others. That is, it should be consistent with your status and in an easy-going style. Try to say proper words and find the opportunity, grasping the discretion and criterion.

(6) The friendship with a big difference in financial or spiritual aspects is generally not long-term friendship. For example, one would be embarrassed if the other always pays for him but he can do nothing back, similarly, the other may feel imbalanced since he has contributed too much, and the friendship would gradually come to an end. Therefore, consider carefully about giving equal opportunities to each other. As the saying goes, "Birds of a feather flock together." Only when people have common ground in the character, temperament, economic conditions, and other aspects of the spiritual world, can they keep long-term friendship with each other. Similar sounds echo one another, and the same odors merge together.

Countries need equality and mutual benefits: people need equality when communicating with each other; men and women need equality between them. These are in accordance with human nature.

3.10 Never spoil yourself

People are all alike. People have few differences in thinking, intelligence and emotion, etc. Although some people are particularly clever, and some are particularly stupid, there're just a small number of them.

Do not spoil yourself. It could be clearly seen that who is good and who is bad, who is kind and who is evil, who is honest and who is cunning in people's frequent contact and interaction. Especially for the people who have certain social experiences, they could know people well from several words, so do not play smart, do not be bad to others, do not be jealous, and do not think that our words and deeds will change others. We should keep in mind that other people can be much smarter than us and they will quickly find out whether our words and deeds are good or evil. One's observation ability is always higher than his performance ability. As the saying goes, "On fluttering wings, the destination it will fly to is known." Playing smart or being cruel to others will immediately receive others' contempt and depreciation face to face or in the heart, which is also cruel for yourself and spoiling yourself. Being kind and honest to others will gain appreciation and praise, which is also good for us to improve personalities. If so, why shouldn't we be good men sincerely? Although there are some "good" crafty men who pretend to be kind, they will expose their own identities one day and all their efforts will be wasted because pretending is pretending and truth is always the truth.

Bad deeds, as well as the good ones, may rebound upon the doer. Do not be angry with those people who are rude, impolite, timeserving, vile and those who do not respect others. Instead, we should laugh at them because their ridiculous behaviors will rebound on them sooner or later. We should revere those people who are kind, polite, and decent because they will win lots of favors and make big progress in their life career.

Don't be confused about the future while doing good deeds. It is worthwhile to be good persons. Try to be nice to others, care about others, and be a good man, which is actually beneficial to yourself. A wise man does not do foolish things, and does not spoil himself.

3.11 Be cautious to everything

Everything in the world needs to be treated cautiously. Carelessness, non-maintenance and tactlessness are not always desirable. Besides, foolish recklessness must be abandoned.

In order to protect the items we use, it is necessary to know their functions and load limits first. The largest bearing capacity per square meter of floor in the building is certain, if exceeded, it will collapse. This may also happen to sofa, chairs, windows and doors. Is there anything in the world that can never be damaged? No, there isn't, even the earth must be carefully protected so as not to be exploded.

Many things are actually tacit in a unit, a family or social communications and once they are exposed in public, they can not be maintained. The relationships between colleagues, husband and wife, parents and children all should be carefully preserved, if not, these relationships will break down. It is almost respectable but not amiable to speak straightforward or tactlessly or not to save others' face. The real person being respected should be easy-going, kind, modest, not serious if necessary, and gentle. He would talk about others' shortcomings in a euphemistic way that can be accepted by others. A person is often hurt or offended when he is in a weak situation, therefore, we should be kind to the weak.

In a society, if the roofs of all houses were removed, what would it be like while overlooked? It is the roof that protects people's dignity and privacy, and everyone could show up holy as usual. However, if people's thinking and behaviors were all exposed to the public, this world would not be the way it is looking like now. For the relationship between two persons, if their thoughts, attitudes and behaviors about each other were all exposed, no good relationship could be maintained, let alone friendships. When you are feeling that you are wronged or unfairly treated, just imagine if what you are thinking and what you are doing are all exposed to the public, I believe you won't have that feeling any more.

Therefore, items would be destroyed without protection; things would be screwed up without prudence. Only being nice to everything in the world is practical. Be nice to others and be strict to ourselves.

3.12 Four one-thirds principle

Reduce one-third while looking at others' faults in case we misunderstand and wrong them; add one-third while looking at others' merits in case we miss something and make them suffer a loss; add one-third while looking at our own shortcomings to avoid self-indulgence; reduce one-third while looking at our own merits in case we get arrogant and lose ourselves.

3.13 Make amends quickly

Errors, mistakes and misunderstandings are inevitable in life, but once they appear, it would be best to make amends quickly. As the saying goes, "A stitch in time saves nine."

"To err is human." Everyone has mistakes, including sages. Mistakes are not terrible; the point is admitting them, correcting them and learning from them to avoid repeating such mistakes. If so, you are nearly a sage.

Subjectively speaking, faults are less severe than mistakes, but not less severe in results. The cause of faults is a shortage of preparation in advance and enough attention. Make amends quickly to avoid repeating faults is the only choice.

Straighten out misunderstandings as soon as possible to prevent chain reactions. More than 90% of the interpersonal conflicts are derived from not clearing up the misunderstandings. We should believe that justice is justice and evil is evil, but there are also many examples of taking justice as evil in history. We should believe that history will give people innocence. But is it too late to wait until later generations justify us? And our misunderstandings are not too severe to let our children to justify us. If we left the misunderstandings for the later generations to justify, they would become a stone dropped into the sea and it would delay our whole lifetime.

3.14 The present is the best

"Time flies like a shuttle." Life is short. People always linger about their younger days. When they are thirty, they say, "If only I were twenty!" When they are forty, they say, "If only I were thirty!" When they are eight-y, they would say, "If only I were seventy!" They are always sighing like this, "It's better to be younger than the real age." But the fact is that the prime time had not been put into good use. One often does not realize the importance of something till it is lost. "For tomorrow, today is always young." We should come to realize the truth: cherish what you already have, cherish today, and cherish every moment, because the present is the best. Take past as the foundation, and grasp every moment today. Yesterday has already passed and tomorrow is still to come, only now is real. Now you may have errors and regrets, it doesn't matter. The present is the best. Just seize the present, and seize what you have. Because only after the psychology and the reality are balanced, can it be the best approach to bring psychological comfort.

The past is "today", today is "today", the future is "today", so we must grasp today; the past is for "tomorrow", today is for "tomorrow", the future is for "tomorrow", so make a good plan for tomorrow; the past is the past, today is the past, the future is the past, so let life run its natural course. When you are feeling sorry about the past, the present is to become the past. Therefore, being infatuated with the past is in vain, and grasping the present is the best.

People have two kinds of age: physiological age and mental age. Physiological age can not be changed, but mental age can be controlled by us. A western saying goes like this, "There is octogenarian child and three-year-old centenarians." The mental age is controlled in our own heart. Being young or old can be converted in our own idea. The present is the best, the present is young forever.

The present situation is your own choice and a historical necessity. If you changed to the other way, it might be worse than what it is now. There is no other choice now; this is the only choice, and also it is the best one. "The present is the best" is the perception of life, is the love of life, is the dependence of happiness.

3.15 Worries are the biggest burdens

Do not worry; worries will swallow up joy, health and happiness. Actually, about 99% of the worries will not happen at all, so relax yourself; about 99% of the dreams will never come true, so don't expect too much; about 99% of the worries are brought by yourself, so do not worry about your imagining troubles.

People spend their whole life time worrying and expecting. From childhood to old age, from day to night, from sobriety to dream, from school to society, from learning to working, from life to career, from children to seniors, from relatives to friends, from marriage to family, from war to peace, from nation to country, from country to world, from nature to humans, how many worries and expectations do we have? But how many worries actually happened? How many dreams have actually been realized? Not too many. About 99% of worries and expectations just meet our mental needs, and bother us. Because of these, we would pay plenty of time and energy, miss lots of opportunities, screw up many jobs, and add more troubles, upsets and sorrows, get diseases, have more conflicts, misunderstandings and ridiculous behaviors ... "Don't be confused about the future while doing good deeds." "Don't ask about the height while climbing." "Let nature take its course." Do not suffer from imaginary fears and do not expect something impractically.

3.16 New things must have new problems

New things must have new problems, no matter they are small new things a single person tries or new things in social development. We should pay great attention and deal with them carefully.

New things are fresh things, which haven't been familiar to or known by the people. That is why they will cause new problems and people don't have preparations.

Be cautious about new things and do the following things: a. To know it as much as possible; b. to forecast and list all the possible problems; c. to prepare preventive actions after good evaluation of their feature and severity.

New things may be good or may be bad; they can be progress or can be regression. New methods must be used to deal with new things.

3.17 Language weapon

Language is one of the most powerful weapons. Language is the commander, action is the performer and a person who is as good as his words is the conqueror.

People use language to communicate. The world will lose connections and become imperfect without language because the world is made up of objects and the relationships between objects.

Battles are commanded by language. If there is no language, there will be no commands.

Work is coordinated by language. If there is no language, there will be no success.

Interpersonal communication needs language. If there is no language, there will be no society.

Every country around the world wants to demand a greater say in the United Nations, so in some sense, language plays a leading role. The person who controls the language will control the world. There is an English proverb which says, "Words command." It means the language can command everything.

Language is a double-edged sword. Language is not only a useful weapon, but a harmful tool. Wounds hurt body, while words hurt heart. Wounds in body are easy to heal, while heartaches are easy to relapse. Language is a useful tool for a good life, so everyone needs to refine it; language is a double-edged sword, so it needs to be controlled by the heart. There are several sayings about language, "What is said cannot be unsaid." "Open your eyes quickly and your mouth slowly, hold back part of what you are about to say." "A kind word is remembered for a long time, but an abusive language hurts the feelings at once." "Saying is more important than planning." People may forget what you've said, but will never forget the impression you gave. You will deeply regret for making such big mistakes which are caused by the words you've said. Choosing words carefully creates success. Confucius[①] said, "One word can make a country prosper or decay."

Praising words are productive force. Men are emotional animals, who firstly desire to be respected and recognized. Things are done by people and it is the prerequisite to arouse their enthusiasm in order to do things well. However, language is the first element to make people have the willingness and enthusiasm to do things. Appropriate words can make it possible for people to change from negativity to positivity, from passivity to initiative and from muddling through their work to working vigorously, which will emancipate and

---

① Confucius, who was born in 551 B. C. and died in 479 B. C., is a great thinker, educator and philosopher well-known around the world. He is the founder of the Confucian school.

develop productive forces. On the contrary, non-appropriate words can hurt people and even throw a wet blanket on their enthusiasm, which will impede the emancipation and development of productivity. There are those who are addicted to tobacco and wine, and there are also persons who are addicted to praising words. People can be divided into three kinds according to the level of addiction to appropriate words. First, the number of those who are indulged in praising words amounts to 70%. Second, those who like appropriate words amounts to 20% and third, those who think appropriate and non-appropriate words have no difference amounts to 10%. Appropriate words, praising words and acceptable words can relax and entertain people's mood, stir up their zeal and arouse their vigor because appropriate words come from the bottom of one's heart. Therefore, it is a golden rule to express serious things with an impression of smile, usual things with a tone of humor, and joking things with an attitude of caution; say appropriate words boldly, say no malicious remarks, say less subtle words and avoid saying reproaching words. Appropriate words are productive force, but it is not enough to say only praising words if the material inspiration or encouragement is neglected.

3.18 Hold the initiative forever

Everyone wants to hold the initiative, because no one is willing to be restricted by others. However, you must leave yourself some leeway if you want to hold the initiative. Just as a proverb says, "The bow should not be pulled fully, and the strength should not be worn out."

In the human communication, under the situation of stalemate, you should make a concession when the opponent is in the inferior position, thus you can keep the initiative forever. On the contrary, if you take the unfair advantage of the situation and do not give face to others, you will make yourself in a passive condition. Therefore, it is important to leave some leeway, that is, by charging you can attack, and by fleeting you can defend, which makes yourself hold the initiative forever.

In the relationship of leading and being led, if you are the leader, you hold the initiative. But if you can not use your power properly and make some mistakes, even some resentment, you will lose your initiative at last.

In the social undertakings, be wise to your investment in the activities of economy, culture and so on, and you will always be in the initiative condition.

In the individual development, you should always pay attention to your body health, because body is the base of keeping initiative. Great ambition is the requirement of initiative, and the consistent efforts are the guarantee of initiative. Only when you are foresighted in seizing the opportunities, and making the correct predication, can you hold the initiative forever.

It is said that "Empty is reality, reality is empty." The punched fist has no strength, while it is powerful when the fist is ready to punch. If you leave some leeway, you will be on the initiative position; otherwise, you will be on the passive position.

Chapter 4 Career

Everyone in this world is longing for career and success. But what is achievement (namely the combination of career and success), and how should we grasp it? This chapter will provide you with a detailed illustration.

4.1 Success lies in the details

The great originates from the commonplace and success lies in the details.

The details show the delicacy, the efforts, the thoughts, and the spirit. When we are helping others, the details embody honesty and duty. For the people who accept help, the details embody their kindness and weakness. When we are giving others things, the details show consideration and love. When we are accepting things from others, the details show our thanks. When we are doing things, the details can promote the success, while in the leisure time, the details can show our interest. When we are communicating with others, the details can help us understand others well. While we are dealing with matters, the details can help us evade unnecessary troubles. For the couple, the details can show great passions. For relatives, the details are symbols of intimate emotions. For elder people, the details can teach others skills. For the young, the details can express the filial piety.

The richest Asian man, Li Ka-shing[①] would often pick up one penny on the ground. This detail embodies his good behavior of economy, and this is one of the right elements of all his success. Meanwhile, the detail reflects one

---

① Li Ka-shing(29 July, 1928—), born in Chaozhou, China, is a Hong Kong business magnate, investor, and philanthropist. He is the richest person of East Asian descent and the eleventh richest person in the world with an estimated wealth of US $33.5 billion.

kind of great thinking and one kind of traditional morality.

Once, an achieved entrepreneur delivered a speech in an economy forum. After the speech, a bank manager walked to him and asked, "Do you need a loan?" He felt a little surprised, but answered, "No, I don't need loans." The bank manager passed him a name card and said, "If you need loans, you can ask me at any time." He felt rather confused and thought that he had only made a speech, why the manager was so sure to loan to him. The bank manager said, "In fact, I don't know you well. I noticed that the shoes you wear may have undergone three years. I'm very surprised that you, as an achieved man today, can still wear such kind of shoes. For this I am confident to loan to you." The detail of austerity is the reflection of massiness in thinking.

Once upon a time, a boss wanted to hire an aid, and there were two candidates. The boss treated the candidates with noodles and gave each of them a full bowl of noodles. One ate as soon as the boss gave him the bowl, but he spilled a lot on the table, while the other ate the noodles very carefully without any spilling. It was obvious that the careful one got the job at last. This is the value of details.

In April, 1930, Yan Xishan and Feng Yuxiang① made an alliance of anti-Chiang Kai-shek and started Zhongyuan War to condemn Chiang Kai-shek. Yan's Army planned to join forces with Feng's Army at the crossing place of Henan Province and Shanxi Province in order to defeat Chiang Kai-shek's Army which was collecting in Henan Province. Unexpectedly, when the secretary of Feng began to make the war plan, he mistook the place of "沁阳(Qinyang)" as "泌阳(Biyang)". Coincidently, there is a real place named "泌阳(Biyang)" in the south of Henan province, about hundreds of miles off "沁阳(Qinyang)". Then Feng's Army went to "泌阳(Biyang)" by mistake and delayed the opportune time of eliminating Chiang Kai-shek's Army, causing the failure of joint operations.

Once, a worker repaired the air conditioner for a family. He wrapped all the line with rubberized fabric, examined all the dangerous parts and changed the broken parts with new ones and sometimes he would return to change if he had no time to spare changing at the first time. The worker only charged the

① Yan Xishan and Feng Yuxiang are both local warlords at the time.

family 200 *yuan* for all the work. How considerate! How heart-warming! The procedures of moving air conditioner are rather simple: removing, moving and resetting. But not everyone can do things precisely and in such details like him, while the details tell everything and make the basis of future success.

The ordinary work of the secretary or office members in the office is doing the cleaning, simple as it is, but can all the staff members think of changing the page of the calendar? Exactly the detail of changing the date page can be an impressive act in daily life.

One enterprise bid for the gratuitous capital for developing the new products. The new product belonged to the high technique, within the scope of gratuitous capital with ready conditions. But because of the mistake of one item, the debt rate spoiled all. The actual debt rate was 27 percent, but was mistaken as 72 percent. And if the debt rate listed beyond 70 percent, the project should be refused. At last, the enterprise was not only devoid of the gratuitous capital but the pre-work turned to nothing.

In one Sino-foreign venture project with closely fulfilled preparation, the Chinese invited foreign businessmen to banquet before signing the contract. During the banquet, the Chinese leader was so hospitable and he frequently invited the foreign cooperators to drink. However, frequently inviting customers to drink is not welcome in western countries. The wine culture in western countries is quite different from the Chinese one. The foreign cooperator got drunk seriously owing to the leader's toasts. The next morning, the foreign cooperator sued in the court for the Chinese leader's intentional hurt, and then the contract got empty. The project failed just because of neglecting the wine culture background. The detail spoiled the whole thing.

There was a company which was hiring employees. The applicants should enter the house through a gate to be interviewed one by one. There was a person waiting at the gate and he opened the door for the interviewees. Some interviewees went into the house hurriedly, neglecting the person at the gate, let alone saying "thanks" to the person, while some interviewees were very polite and said "thanks" to the person. It was obvious that the company would choose the ones who were polite enough.

A warehouseman was wanted in a company, and the recruit place was in the warehouse. On the floor, pieces of paper and sundries scattered. Most of the employees concentrated on the recruit worker and only one employee bent to pick up all the "litters" and put them away. The result was that this man got the job.

Before starting a major event, the general procedures of it are exemplified in the process of planning. But actually the success lies in the details, as the omission of small detail may cause the failure of the whole plan. Just as the saying goes, "A small leak will sink a great ship."

To start a major event, we should make sure the general steps are done and should be done in this way. But can the details always be fully considered and done well? The results will be different. If the details are done well, people will be touched and feel grateful, and then the result is successful, while if the details are done badly, people will turn to be unsatisfied and even doubt the honesty, not to mention success.

The reasons for failure are not the improper plan, inadequate conditions, incapability and having not made efforts, but the omission of details. What a pity!

Five hundred years ago, the Spanish Queen told Columbus who was setting about, "Angels are in imagination, while the devils are in details." Carnegie once said, "A man who does not care about details will never make success."

When all the people strive for success, caring about details becomes the critical measure.

The details show efforts made behind them and efforts lying in non-stop practice. The details may appear inborn or be practiced through acquired efforts. The man who was bitten once would care more than before. The details can be recognized and done well only after the man has owned abundant experiences. The prince in the palace even can not brush his teeth when he grows up, how could he think of the details of brushing? The children who were brought up in poor families have to support the family in their young years, and often this kind of kids knows more details of life and possesses abundant experiences. Like the exercises which the students do, the common practice should be treated as one examination, and then the abilities will be accumulated day by day and month by month. Experience is precious, whether it is successful or not. The details are key elements, no matter for others or for yourself.

4.2 Drawbacks are neighbors of success

The way to success is full of frustrations. Edison has 2,000 inventions and over 1,300 patents in his whole life. However, he had met numerous drawbacks and failures. Yuan Longping①, father of Hybrid Rice Breeding, who has spent over twenty years in the rice field in order to study the new cross-bred breed, worn out scores of pairs of shoes and several motorbikes. After numerous obstacles and failures, Yuan has made great achievements. "Honing gives a sharp edge to a sword, and bitter cold adds keen fragrance to plum blossom." No one can succeed easily.

You must walk your own way, you must understand the principles by yourself, and you must experience your own life. You should experience your own life as it is yours. If you try to do something, you may succeed or fail. But if you are afraid of failure and dare not try, you are deprived of the chances of success. George Bernard Shaw said, "A life spent making mistakes is not only more honorable, but more useful than a life spent doing nothing." Failure is the mother of success. No drawbacks, no success. Drawbacks are the neighbors of success. If the drawback comes, can success be far behind? Someone says that failure may cause serious and even lethal strike. But if we prepare two results of being good and bad previously, we can solve problems effectively after the thing happens without worse loss and drawbacks. "The precious point of the strong lies in restarting after failure, while the pity point of the weak is the decadence after failure." It's more important for you to trust yourself than believe in others. My daughter, Wang Fei ever said, "I succeeded at the key time by myself, but failed when I depended on others." The basis of life and important career of life is to enrich and temper yourself. When you have chances, you should make contribution to your state and people, but if you do not own such chances, you can work for yourself. "Where there is a will, there is a way." It can't be too excessive to estimate the unfairness in the world, but the righteousness can never be destroyed by suppression, as there is always a fair public opinion.

Courage and perseverance are essential for success. Though the thoughts

---

① Yuan Longping is a famous scientist, and an expert of Chinese Hybrid Rice Breeding.

are important for success, but courage and perseverance weigh a lot more than thoughts. Most people have the exciting thoughts and ideas, but those who have courage to carry the ideas out are less. Edison said that success is 99% of perspiration and 1% of aspiration. Everyone has the one percent aspiration, but not everyone is willing to make 99 percent effort to realize their dreams. To do well with a thing, you should possess the spirit of Kua Fu① who were chasing after the sun; you should have the perseverance of Jing Wei②, who is a symbol of dogged determination; you should practice the strong willpower of Yu Gong, who determined to move the mountain standing in front of his house; and you should have the diligence of "rise up upon hearing the crow of a rooster and practice with the sword". Though these characters are from different stories, they share the same point in common. That is, all their ways are full of drawbacks. Laziness causes poverty, while diligence creates wealth. The natural law deserts laziness. Inertness is a kind of tendency and habit, different from the laziness in common sense. Those who have inertness are impossible to own hope when they face the music. Everyone has inertness, but if you let it go and get into the habit of it, you'll spoil yourself totally. Inertness turns to badness, then badness to harm, and eventually harm to yourself and others. Inertness eats up your own happiness, destroys others' benefits and breaks the public rules. Thus inertness can be taken as one of human beings' enemies which can bring about more obstacles.

Knowing others are beyond you, you'll always be a winner in the society. Lao Zi has ever said, "To understand others is to have knowledge; to understand oneself is to be illumined. To defeat others is to have strength; to defeat oneself is to be strengthened." To understand others and to win oneself are more important in the saying. The premise of the former one is to differentiate others, while the premise of the later is to understand yourself. As the proverb goes, "One's most difficult enemy is oneself." If man can understand others and himself, and can win himself, he'll be a winner in the world. But how can you be a winner of yourself? First of all, you should understand yourself and possess confidence. Second, you should have the controlling force of yourself. Third, you should have perseverance. It is obvious that the premise of self-understanding and confidence is that you've really experienced

---

① The story of Kua Fu is one of the ancient China's myths and legends.

② The stories of Jing Wei and Yu Gong are the ancient China's myths and legends.

a lot and have quite a number of capitals by material or spirit. Self control is the result of broad mind and self-cultivation, and it is the radical strength of winning yourself. Perseverance is the result of experiences and practice, and it is the basic element of success. You can reduce the drawbacks by winning yourself.

Wonder comes from misfortunes, while greatness grows out of difficulties. Chen Jiru① said, "Celebrity always lies in the poor days, while failure often visits in the famous sections." A man's potential is marvelous. However, most people can not realize it, because the potential can only be shown by stimulation and the dilemma is the right stimulation. As long as you are brave enough to challenge yourself, you can realize yourself and even release your giant potential; as long as you are not afraid of undergoing obstacles, you can make great achievements. The more difficult the situation is, the greater achievements you will make. As long as you don't succumb to the environment, you can change your surroundings. The environment makes the man. The man who is good at making full use of things can make success, while the man who neglects the current affairs will be deserted. The clean and proud pine comes from chilly weather, and the hero turns up in difficult times. Good luck can be gained through insights and continued diligence. Don't trust luck, but trust yourself. The great men don't believe in luck. They can seize every chance, no matter how little the chance is; they can create chance, no matter how difficult the process is.

There is a record in the book named *Biography of Qin Emperor* that Lü Buwei said some impressing words to Ying Zheng, the name of Qin Emperor, which played an important part in the politics at that time. "The genius and the great are both lonely, no matter they have undergone ups or downs; they are always deserted by the ordinary beings. Have you ever noticed the scene that when the red-crowned crane fell in the flock of hens and was bitten by them? Can you imagine the dragon that was trapped in the shallow water and played by the fishes and shrimps? Don't get surprised at these ordinary phenomena as they are different in nature. Once the crane flies, the hens are still seeking food in the field. And when the dragon flies and makes great changes in the sky, can you find the fishes at that time? You can never find them again!"

---

① Chen Jiru (1558—1639) is a famous litterateur and artist in the Ming Dynasty.

4.3 Kung fu is the accumulation of painstaking

Kung fu is the attainment and capability in one field. Once the actions, gestures, strength, and skills are accumulated, they can turn to kung fu.

Once upon a time, there was a poor guy who wanted to learn kung fu from someone. As he was so simple and honest, the master was not willing to teach him and said, "Go home and practice moving the stone with sticks." The simple guy believed in his master, treating his words as the real knowledge of kung fu and practiced this action every day. At the very beginning, the boy could not move even the little stone well with sticks, but gradually, he could move the huge stone with sticks easily. One day, he heard that there was a competition of kung fu in the marketplace and went to have a look. When he went there, he saw his teacher was fighting with the kung fu holder, but his master was defeated soon.

He went to his master and said, "Master, I'm coming."

"Who are you?" The master got a little confused.

"You've forgotten me? I'm one of your apprentices. You asked me to practice moving stones with sticks at home and I've made it."

"Go home! Quickly! If the winner gets to know you are my apprentice, he will kill you."

"Master. I have come to win back your face. Wait and see."

He went straightly to the stage.

The kung fu winner shouted, "Guy, what's your name?"

"The man who was fighting with you just now is my master. I'll fight with you again." The young guy said.

"Come on, silly guy!" The holder made his pose.

But the young guy stuck the "winner" highly in the sky with a stick at a time.

Another young man also learned kung fu from his master. His master asked him to twist ropes every day. The ropes thickened day by day and the young man's arms were strengthened more and more. When he was in the fight field, he defeated his enemy through this kind of strength to safeguard his homeland.

There was also a person who kept carrying the calf when the calf was born. The weightier the cow became, the more powerful the young man got.

When the little calf became a big cow, the man could carry the cow easily. When he competed with others, even the strongest man would be thrown out easily by him.

The Shaolin Temple is the noble place of numerous kung fu fans. The Shaolin kung fu is of extensive knowledge and profound scholarship, having learned wildly from other martial arts with a long history of kung fu soul. It is said that to practice kung fu in Shaolin Temple is quite a difficult thing. The kung fu coach monk will not let you take a rest until you are tired to faint. Before then, you must persist in practicing and can not be lazy for a while.

4.4 Expectancy can gain outstanding achievements, while indifference gets no present result

The scenic spots which are far away are attractive, while the close scenery cannot be interesting. Maybe everyone has this experience that we are not interested in the close sceneries but take interest in the sceneries far away. The reason is taken for granted that the sceneries which are near can be seen at any time. But some people will not see the near sceneries in their lifetime as they do not have such expectancy. No expectancy, no results.

The expected ones are the best. It's just like looking for a partner in marriage. When you have expectancy, a goal, you'll take actions to court him or her. After long term wooing, you realize your dream at last. But if you do not have such expectancy, you will make no plan, not to mention realizing your dream. Even though others court you, and you two get together at last, you cannot get true love, just because people always do not cherish the things that come by chance. What you get at last is not what you really want to gain.

Expectancy is the premise of making success. Those who have the conditions of making outstanding achievements cannot succeed without desire and expectancy, while those who have no conditions of making achievements may succeed because of his strong desire and willpower to create conditions to succeed. Just as Kahlil Gibran in Lebanon says, "Will is half life, while indifference is half death."

There was a story of expectancy. In a place, an earthquake took place. Three days later, a man was still seeking for his wife in the remains of the building. Many well-wishers told him to give up, "It's too late." Nevertheless, the man persisted in seeking. All is quiet in the dead night and he was still busy turning over the remains and digging. Suddenly he heard the weak voice of "help". From a triangle cave formed by the clasping stairs and broken wall, he found his wife and rescued her finally. Where there is hope, there is possibility of realizing the dream. Where there is no hope, there is no expectancy at all.

4.5 Career is pursued

Living in this society, especially for the young who are just entering the society, when they face the complex world and crowd, they'll feel small and weak, for they have no resources and power. What do they depend on? Snatching or robbing? These methods are discouraged as they are illegal, then only one approach is left: by pursuit.

Liu Bei[①] visited the cottage thrice in succession to call on Zhuge Liang[②] repeatedly to enlist his help. He had been pursuing all his life and got the territory at last. Cao Cao eagerly sought after men of worth and ability, and finally made outstanding achievements.

In the construction of our country, the service industry pursues customers; the production industry pursues providing and selling; the country pursues prosperity; the people pursue harmony. We pursue rainy days during the drought, and fine days during water log.

It's more common to pursue in social life. You pursue marriage when you are single. After getting married, you pursue children. When you are healthy, you pursue happiness. When you fall ill, you pursue medical advice. When you are jobless, you pursue jobs. When you have a job, you pursue wealth. When you do business, you pursue prosperity. When you are in office, you pursue promotion. When you begin an undertaking, you pursue talents; during the development of your career, you are courteous to the wise and condescending to the scholarly. When you are in danger, you pursue talents for help. When you know nothing, you pursue knowledge. When you are puzzled, you pursue reason. When you make mistakes, you pursue for-

① Liu Bei(161—223) is the first emperor of Shu-Han and a great statesman.

② Zhuge Liang(181—234) is the prime minister of Shu-Han, one of the greatest strategists, statesman, engineer and scholar.

giveness, and when you are right, you pursue encouragement.

In boyish years, people pursue; in adult years, they pursue; and in their elder years, they still pursue. They pursue home and outside; they pursue in difficulties and in happiness. Pursuit is a life-long engineering for everyone. Do not be a miser on pursuit and do not hesitate to pursuit. Everyone needs pursuit but pursues a bit less luxury.

In a nutshell, no matter what you pursue, you should have something on the ball. "The premise of striking irons is to be hardened by yourself."

4.6 Career is connected with kindness

One should be kind to begin one's career. Those who are spiteful cannot make great achievements.

As the career is made for the people and society, the spiteful cannot do good to others, thus they cannot make great achievements. If you want to make great achievements, you should be considerate to others; you should relate yourself to the career of people and the future of human. Career is the overt embodiment of the inner heart, thus we should cultivate our minds before we begin our career.

4.7 Hold your base tightly

It is changeable in prosperity and decline is capricious in rise and fall, so it is important for you to hold your base tightly. It turns out that when it is dark in the east, it is light in the west; after it is light in the west first, it will be light in the east next. We can get the favorable side at one time and unfavorable side at the other time. "Everyone will have his turn, and I may also be an emperor tomorrow." Nowadays everyone lives a fast life and the change can be seen in a shorter period, a period of ten years or much shorter than that.

When there is a career that almost no workers are willing to do, it is evident that this career will turn to be a hot career soon. That is because everything develops between balanced and unbalanced state. All the careers are necessary and balanced in the society, and they need to keep balance.

Not long before, no one was willing to do funeral and interment, but now it is so profitable that the profit rate can reach 1000%. Not long before, the peddlers were looked down upon, but now they are the symbols of wealth. Not long before, working in the foreign trade corporation was admirable, but now these corporations suffer from great depression. Not long before, the bankers enjoyed favorable salary, but now it demands a lot on the workers with task goal in a fixed period. Not long before, the job in the post office was a classy one, but now it can only keep the wolf from the door.

Some people changed his age to a younger one, but when other contemporaries meet the preferential policy of retirement and enjoy the pension at home, he cannot get the pension because of the younger age. Some people worked in the state-owned unit, but he asked the leader to change his identity of worker to cadre. However, later the state-owned unit changed into enterprise. According to the principles, if he was a worker, he could retire in his age and get favorable pension, but as a cadre, if he does not reach the age of retirement, he can only get fewer pensions than as a worker. Some people always admired profitable jobs and positions and changed jobs frequently, but did they get a satisfied job at last? Though he might get a job finally, he must start from the very beginning and face new and difficult pressure. However, his fellow colleague who was still staying in the previous position, got more chances to be promoted. Even the staying ones were lagged behind of the job-hopping ones, they got the leisure and relaxing life. For those busy people, they often make use of the spare time to do more things, managing to do both the principal work and sideline occupation successfully. It is also wise for you to change jobs if you cannot move forward anymore in this position. "The tree will be dead by moving it, while the human will be alive through changing his surroundings." It may make the difference.

As long as you can use your head, every career may be a chance for you. Every dog has its day. As the saying goes, "A lucky dog needs no haste; an unlucky dog runs heart broken." If you are unlucky, the results will be all the same whether you're busy or not. The meaning on the first level is to tell people not to be over busy. And on the second level, it is to say the state of existing. It's often the fact that the lazy one gets happiness at last, while the busy one is devoid of happiness because of over diligence. It is a wise way to hold your own base tightly as mentioned above.

4.8 Right time, right place, good human relations, and calm mind

A saying from ancient time goes like this, "Right time, right place and good human relations" are the three key elements of success. The three elements are indeed important for success. Without them, one cannot succeed. However, even if one owns the three key elements of success, he may not succeed as he may lack another element, which is the calm mind. The calm mind is also a critical element in making success.

The calm mind refers to that the mind is fully at ease, unperturbed and fair, free from pain and emotional upheaval. He who has a calm mind looks down upon the utility, fully understands the life, and is calm to himself and others and then the whole world. While meeting difficulties and drawbacks, he should not feel frightful, not abuse himself excessively and not be narrow-minded. When he feels fair in mind, it equals to gains in fact. When he faces benefit, he should not be greedy and gainful, and should not try to rob the wealth in all ways. Even if he gained something in all mean and extremely vicious ways, he could not get a calm mind and actually he must have lost something. The right time, right place, good human relations and the calm mind are inseparable parts of making success. There are three states of mind of life: positive, negative and mild. The mild mind is the basis of positive mind and the balanced code of negative mind.

Frame of mind produces state of mind and decides on feelings and actions. The appreciation and assessment of the world originates from frame of mind. If you are in good mood, the common flower can be a beautiful one; but if you are in a blue mood, the most beautiful flower can not attract your interest. In a word, the flower of mind is the most beautiful. The mild mind brings about the mild surroundings; the restless mind can cause unsafe surroundings. The main question of methods is the attitude, and the inherent question of attitude is the frame of mind. Narrow, spiteful and cursed frame of mind can not bring about good attitude and proper methods.

The mind is not always tame as we often lose the right time, waste the right place, spoil the good human relations and destroy the whole expectancy, thus we can say that we are the giant enemy of ourselves on the way to success. Man can defeat everything on the condition that he defeats himself. What aspect of himself should he defeat? The answer is to defeat his state of mind. Keep a mild mind, and you can enjoy long life, harmonious relationship and successful business, thus mild mind is necessary for successful life. We should treat it equally as the three key elements as we've mentioned before. The right time, right place and good human relations are objective conditions, while mild mind is the subjective condition. Only when the subjective conditions are combined with the objective ones, can the man make good use of objective conditions and make success which is admitted by himself and others.

4.9 He who has the cards in his hands will win the game

The success lies in the previous correct inference on things and effective control of relative elements. He who has the cards in his hands will win the game. Previous planning is the key of winning. Comprehensive plan makes less trouble later.

When you begin the career, you should forecast the result at first. The giant benefit of enterprise is not the rolling development of the present asset, but the correct forecast of more profitable future. The famous investor—Warren Buffett is the great practicer of value investing theory. He begins with analysis of enterprise's financial statement, studies enterprise's management and long-term competitive strength, and appraises the intrinsic value on the above bases.

Buy the odds-on stock for good prediction. A little move in the stock market will cause great vibration. When the accident takes place, the stock has already been sold out and it seems that nothing has happened.

The key point of driving is forecast. When you drive, you should forecast all correctly, including the front and back directions and the aspect of reversing, then you can move your car easily and flexibly.

The precious weapon of defeating the enemy lies in the odds-on. Fighting with the enemy, you should have a comprehensive understanding of the enemy and yourself, making correct forecast and calculation, and then you can defeat the enemy in proper means.

To suppress the revolt, you need forecast. You should be prepared for danger in times of safety, make adequate preparation, including arranging spies. And when revolt comes, you can put yourself forward and stop it easily.

Think more for next step while in this step, and things will go well through full forecast. As the saying goes, "You cannot eat and dress up to poverty, but inadequate forecast can." Sun Zi, the most famous militarist in ancient China said, "More forecast, more winners. Less forecast, less winners. Let alone no forecast?" Chen Danran in the Qing Dynasty said, "Those who do not plan for all ages cannot plan for a little while; those who do not plan for the whole cannot plan for a small area."

4.10 How to make success

Success in life is a complex engineering. The essential of success lies in your own feelings. Success is the mixture of many aspects.

4.10.1 Concept of success

To begin an undertaking is the main content of life; success is the supreme goal of life. Beginning an undertaking and success are not narrowly defined, but generally defined. Every word and behavior of human beings can be a beginning of an undertaking and every standing point of human is the goal of success. Success is the overall success in all aspects. For example, family is a standing point. If you are capable of tidying up the house and keeping it comfortable and clean, and all family members can get on with each other harmoniously and enjoy the happiness in the family, then you are making a kind of success. Couple relationship is a standing point. If the two of the couple respect each other and care each other, depend on each other all the time, then what exists between them turns to love not hate, happiness not disaster, joy not sadness, benefit not loss. They are making a kind of success. Child is a standing point and the hope of life. If he or she can be brought up to be a hopeful person for the family and the society, then the parents are making a kind of success. Neighbor is a standing point. If you can make a harmonious neighborhood, you are making a kind of success. Kin is a standing point. If you can form a custom of respecting the elders and caring the youngers, making the kin members live harmoniously, you are making a kind of success. Relatives are a standing point. If you can keep in touch with them frequently and enjoy intimacy with them, you are making a kind of success. As for friendship, the success is sharing the difficulties and happiness together with them. As for colleagues, the success is getting on well with them without discord, and building up the atmosphere of comfort. Work is a standing point, and the success is to make progress in public and to stand in a right position. As for learning, if you can stick to the motto of "Never too old to learn" and get enormous joy and interest from learning, then you will make a kind of success. The success of entertainment is relaxing both mind and body happily and doing some good adjustment and rest for yourself. The success of sleep is being capable of sleeping sound and bringing about ample energy for the next day. The success of diet is having a balanced diet and being free from any disease caused by imbalanced diet. The success of wearing is being expert in matching clothes, neither luxury nor dowdy, neither wasteful nor costive, but always in appropriate clothes.

In all, any time and behavior of life are all making a man. The only success is the success in all aspects. The biggest undertaking is to enjoy a healthy body; the biggest success is to lead a life comfortably. However, before you start to do anything, you should consider your homeland and the whole world at first, thinking more for the whole human beings, thus you can make yourself a real successful man at last.

4.10.2 Conditions for success

There are many conditions for making success, and the necessary conditions are in the following:

(1) To keep a positive mind. The key element of success is that you should keep a positive mind. Those who always keep a positive mind are destined to succeed at last. But those who are negative often let go the chances of success.

(2) To set a feasible goal. No goals, no success. People with the infeasible goal cannot make success.

(3) To make permanent efforts. Put your goal into practice. Nearly everyone has made a blueprint and set goals for himself, but it is a common phenomenon that less people carry them out. All the things in the world need cost, so does success. If you want to make success, you should make great efforts. The efforts include basic conditions, strategies of executing, perseverance, etc. Success needs strong will and patience. The secret of success is self-improvement, no matter for a person, an enterprise, and even a country. The important point of success is to have the self-knowledge and adopt others'

good points and avoid their shortcomings.

You should make full use of the best subjective and objective conditions to do the most important things. Success lies in the constant efforts and continuous progress. To make progress is to overcome the obstacles, while to retrogress is to reset new obstacles. You are the host of success and failure. No one can help you except yourself in the world.

Success is the product of lofty ideal and bold practice. It's the mixture of strategist and doer. Once upon a time, there was a wise professor who lived next to an illiterate neighbor. Although they had very different status, they shared the same goal—to get rich as quickly as possible. The wise professor talked about his understanding of how to make wealth every day, while the illiterate neighbor listened to him carefully by side. The illiterate neighbor admired professor's knowledge and wisdom, taking actions according to the professor's wealth assumption.

Years later, the illiterate neighbor became a real millionaire, while the professor still talked about experience of getting rich with empty pockets every day without any practical actions. Just as the people said, "Professor, professor, you are only a talker."

Though the illiterate built up his wealth base, he could not have further development in business. After consulting with professor, they came to agreement: cooperative business operation. Under the combination of wisdom and actions, the two persons became the richest in the financial circle immediately.

(4) To innovate continuously. The capability of innovation comes from your own chronic thinking and social experience. The origin of innovation is from one's chronic thinking and different ideas from others. Changes of objects can be divided into three sections: mutation, shading, and qualitative change, so do changes in the process of innovation. In the beginning you have novel ideas and feel joyful, then you may be slow in the carrying out, but the following will be qualitative change and success. Patience and perseverance are critical for success. The time you meet frustration is the time you get close to the achievement and the time you can hope for the coming of success. "Success often exists in the more persisting efforts." There are ten ways consisting of one right way and nine wrong ways. If you've tried nine wrong ways, then you can find the right way soon, and this right way is the way to success. There are nine wrong ways with only one right way, and this right way is just under your foot.

(5) Extent of satisfaction. The satisfaction is your own feeling of satisfaction. The success that you can truly sense is the real success.

In conclusion, to make success, one should choose the right goal, carry the biggest hope, make the most appropriate efforts, stimulate the most enthusiasm, walk in the most correct way, take the most effective measures, and prepare for the worst result.

4.10.3 Success ratio

Success or failure is up to one's feeling. The success which is admitted by public may not be success in his own eye, and he may commit suicide the next day because he feels that he is rather unsuccessful and even the biggest failure. Sense of achievements is the self feeling, while the life value is ratified by the public objectively, so the sense of achievements is one's own and the appraisements belong to the public. Self value is different from life value. To realize the self value is to fulfill one's own will, while to fulfill life value is to make contributions to the society to the largest extent.

Everyone has his successes and failures. Whether one is successful or not is measured by one's own standard of success to calculate the success ratio. You can list the main courses of life and score for yourself. The full mark of each course is 10 points; the passing mark is 6 points. The total of all courses' points are divided by the number of courses. If the result is beyond 6 points, then you are successful generally.

You can set the main courses by yourself. For example: job achievements, state of marriage, education of children, relationship of kin, relationship of neighbors, relationship of colleagues, relationship of friends, state of health, happiness degree of life, accumulation of wealth, etc. There are ten main courses, and you can omit the less important courses by yourself.

This appraisement can fall into two kinds. The first one is public style, that is, to list many courses which are publicly recognized. The second one is unique style, that is, to list a few courses which are unique and stylistic. But it doesn't affect the result as success is up to one's own feeling.

4.11 Cultivate yourself before doing things

Cultivate yourself before doing things. Be a good man first and you can succeed and do good things. Being a good man is not for others but for you.

Being a man is a complex and serious question. Socrates said, "The final aim of all knowledge is to answer how to live a life." The nature of how to live a life is how to be a man. The doctrines of being a man in Confucianism are "Humaneness, Righteousness, Proprieties, Wisdom and Honesty". The root of being a good man is to develop good morality. Morality is all-conquering in the world, and morality advances victoriously everywhere. The way to test a man's morality is to observe how he treats those who are relatively disadvantaged groups. The principle of being a good man is to be honest, and the basis of a man's standing on the earth is also honesty. Those who have honesty often succeed, while those who lack honesty always fail. Here is a story. Once, an old King wanted to adopt an honest child as his son. He distributed every participant child some seeds and asked them to grow them. When the limited day came, all the hopeful children held their flowerpots which were full of flowers, waiting for the King's check-up. But the King chose the child who carried an empty flowerpot as his son. You may guess why. The reason is that the seeds which were allocated by the King were all boiled. They can not grow at all.

Effect will never disappear. All the behaviors of one man, including his thoughts, even dreaming can accelerate his body to excrete bad or good substance, which can influence his body and mind directly. If one has done a bad thing, he would be in fear and trembling all the time and thus his body and mind would be harmed. If he has done a good deed, he would be bright in mind and joyful in his heart, which is good for his health. Being a good man, he can sleep well all the time without worries. Good effect accelerates success, while bad one destroys the life. Man should focus on self-cultivation, being calm but not dumb, careful but not indifferent, decisive but not reckless, contemplated but not indecisive, confident but not too proud, hurried but not worried, busy but not messy.

The public eyes are microscopes and magnifiers; they can make out your spiritual practice and maximize your shortcomings and faults. No words and behaviors can escape from public comments. Thus being a bad man cannot be approved by the public, let alone succeeding in doing things and realizing self-value.

Chapter 5 Family

Family is the field where everyone can build their base. If we can pray for blessings, what can we ask from the world? An eternal lasting enterprise will start here and we take the cause as our base. On hearing your words, we can find the best ways of solving problems.

5.1 The couple lot

In modern time, the relationship between husband and wife pursues equality, harmony and happiness. The author has drawn some conclusions from the couple life and observation of the marriage, hoping to have a mutual discussion with the readers.

5.1.1 The trilogy of couple life

(1) Period of running-in (the first ten years). The mutual personalities and shortcomings and other improper aspects will be exposed and form the friction in this period.

(2) Period of adaptation (another ten years). After the friction, the couple have learned to adapt and compromise to each other.

(3) Period of harmony (the following years). The couple have undergone the period of adapting and sensed the real meaning of life in their 40s, knowing that harmony and interdependence are the real needs of each other.

But for the wise couple, the front two periods will be shorter and they will enter the third period more quickly.

5.1.2 Why do the couples become quarrelsome lovers after marriage?

(1) When they fell in love with each other, they did not make adequate estimation of the difficulties and drawbacks of future marriage.

(2) The attraction before marriage is eros, while after marriage it's the relation of emotions. They feel that after marriage they've gotten what they desire in the period of wooing, taking less interest in eros, and adding the thinner base on emotions; they become less concerned for each other.

(3) The emotions between each other need to be deepened and developed. No progress means regression.

(4) "Beauty is in the eyes of the beholder." The couples after marriage often look for faults. And the faults of each other begin to be uncovered.

(5) After marriage, the social relations which the couples need to face become larger and more, bringing about difficulties to each other.

(6) The friction leads to change. If the friction cannot be transferred to adaptation, the marriage may be internecine.

5.1.3 Big men and little women

If a big man marries with a little woman, they are a pair of mandarin ducks (a kind of birds which symbolize love), while a big woman and a little man are a couple of quarrelsome lovers.

Big men are big in appearance but little in mind, while little women little in appearance but big in mind;

Big men are big outside but little in the house, while little women little outside but big in the house;

Big men are big on face but little in practical benefit, while little women little on face but big in practical benefit.

A real man is broad-minded and tolerable, while a virtuous lady is self-restrained and not sharp-tongued.

5.1.4 Three "Should" and three "Not" between the couple

(1) One should give face to the other side, not make the other side lose face. "Face is as important to the man as the bark is to the tree." Face and dignity are of special importance, especially for Chinese. The couple should respect each other and be courteous to each other. One should give the other side adequate face but not let the other side lose face before others. Do not talk about his or her shortcomings and do not blame him or her before others. As the saying goes, "You can teach your children in public, but never your wife." In fact, teaching children should not be done in public, either, no matter what age he or she is in. You should consider his or her feelings on respect

and face. Listening to others' words face to face and teaching wife and children at home are available.

(2) One should adapt to but not violate the other side. "Adaptation" is the natural law, the science, the discipline of existence and development for all objects. On the "bad" habits of the other side, you should not press him or her to correct them, but adapt to and guide these habits to a proper degree. Since you have chosen each other, you should accept him or her as the whole of this person.

The methods of living together can be different for different spouses. Different people have different tempers. Some people prefer the favorable words, and casual and easy-going style. If you violate these, even more beautiful clothes, better diet and more considerate care are of no avail, especially when the other side is in the surroundings of long-term cursed words from you.

Someone puts eating and drinking in the first place. As long as delicious wine and dishes are provided, he or she will not feel uncomfortable even he is cursed.

Someone is fond of dressing up. As long as fine clothes are provided, he or she will mind nothing else.

Someone only cares the money that the other brings home.

Someone wants the other side to stay with him or her all the time.

Someone wants the other side to listen to him or her all the time.

Someone wants the other side to be inner-directed and determined.

All in all, the happiness of the couples lie in inspecting each other carefully, finding out the principles and personalities to adapt to, and pleasing each other.

To be hen-pecked husband is to love wife. Why should one man be henpecked? Here are three reasons. First of all, he is afraid of risking the wife's anger. Second, he has shortcomings himself. Third, the last but primary reason is love. There is no fear without any love. What does a man pursue? The answer is that he is pursuing his wife's joy, a harmonious and smooth family, and the happy life.

The hen-pecked husband is to let the wife have the right hand up. Why should man let his wife have right hand up? There are also three reasons. First of all, women are more delicate. Second, he should take the whole situation into account. Third, he does so because of love.

To be hen-pecked husband has three results. First, the wife will be happy. Second, the husband will enjoy a harmonious family. Third, his life career will be prosperous. If the husband has broad mind and concern for the overall interests, has a heart full of mercy and charities, then the family is bound to be harmonious and the family members will be living a happy life and managing prosperous cause.

Henpecking is the typical way of "adaptation".

(3) One should be tolerant but not critical. Different people live in different environments, and are accompanied by different backgrounds, experiences and feelings. They may be quite different on the ideas, minds and ways of solving problems. Differences will cause contradictions, and then you both need coordination and understand each other. It is also like this between couples. The couple need to adapt to and tolerate each other.

The most difficult thing in the world is understanding and tolerance. You broke a bowl; I left my key at home; you are quite careful, while I am a little careless. Why should we spare our time to be a hairsplitter? No matter good things or not, they are all the content of life. This is life and we should live in this way. Besides, different people have different tempers. Maybe you dislike the way others behave, and others may feel repugnant for your verbosity. Why should the people fuss over trifles and change others in your own standard? Everyone should know that it is hard to change a person. The couple should adapt to each other, not long for changing one another, for adaptation is the best solution of discord. You can tolerate others, but why can't you tolerate your family members? Perhaps this phenomenon has two reasons. First, the two are too intimate, so they do not care and respect each other much. Second, the two are the beneficial community, asking too much of each other and taking that the other should prepare well for the family without being reminded. Therefore, the mistakes exist in two aspects: one is that one hurts the other's dignity of personality, which is seen as an impolite behavior; the other is that they ask too much of each other in a rather simple and rude way. If you want to make things satisfying to yourself, you should do them by yourself. However, sometimes even you do them by yourself, you may not get satisfied.

One should be broad-minded but not mean. The couple should not be hairsplitters but be tolerant; not blame but tolerate each other; not doubt but believe in each other; not fuss about trifles but be tolerant; not be rude but be polite; not be suspicious but be trustful and especially be self-confident.

The condition between couples is like flying kites. You dare to fly the kite in the wind freely, and you can also grasp the line in your hands and pull it back in time. The line is your self-confidence, quality and glamour.

5.1.5 Strategies of keeping good relations between couples

Principles of the couples are as follows:

(1) To avoid conflicts. Do not blame the other in the process of his or her making mistakes, especially not in the time when he or she gets angry. The two of the couple should calm down, and do not pour the oil on the fire.

(2) The two of the couple should adapt to each other by tolerating different ways of solving problems.

(3) One should be amiable and give enough respect to your spouse's opinions. Women's chatter is next to disloyalty.

(4) The two of the couple should care each other and be considerate. Indifference between couples means betrayal.

(5) The two of the couple should protect dignity of each other and give each other adequate face before others.

(6) The two of the couple should be courteous to each other all the time.

In a word, you should adapt to the world and society in a natural way. Besides, the couples need more adaptation. You adapt to me and I adapt to you, thus happiness will occur naturally. There is nothing much difficult to be solved in a family, as about 99 percent are trifles. For a noble man, a well bred man, an outstanding man, a real man, will he fuss about the trifles? As long as you keep a contented and adapted mind, you can make the couple relations and things successful. Considering from angles of emotions, love and life basis, the couples should and have to adapt to each other. The secret of maintaining good couple relations is only one word—adaptation.

Emotions are the ligament of marriage. Where are the couples from? How do the couple and a family come into being? They are brought together by emotions and love, so the maintaining and development of couple relations depend on emotions. If the emotions are developed, protected and respected, then all the problems can find their solutions and the couple will be harmonious and happy. There is no other effective ways except this. When the couple have emotions between each other, money and object will be brought home straightly, as they always carry each other in heart and care each other in a natural way. The story below can give us some implications.

> A shepherd walked forward slowly. Following the shepherd was a goat. Though the goat was not tied, it followed the shepherd and stayed close to him.
>
> A philosopher saw this, got surprised and asked, "You do not tie the goat, why does it stay so close to you?" The shepherd answered, "It is not the line that ties it, but care and love."
>
> The shepherd's words impressed the philosopher most. He felt that the one that can maintain relations between people is not the tangible constraint, but the intangible care and love.

This is called "Effect of Goat" between couples.

It often happens between couples that the husband is fond of dumplings, and the wife says, "I will spare time to prepare them for you." The husband says amiably, "But it really troubles, and we can eat casually." The result is that these words make the wife more determined and vigorous to make dumplings. When they have lunch, the house was full of smell of dumplings, and this is effect of love.

No rigid controls. It is love that makes the two of the couple together. True love is contribution, care, forgiveness, and also wishes of mutual happiness and joy. Since it is the truth of happiness, why shouldn't one be broad-minded and fight for love? However, someone controls and constraints his or her spouse after marriage and even before marriage, then they become mortgages of each other due to love. This is not love, but deprivation of personal liberty, deprival of civil rights, kidnapping, the extreme possession and greed, being against the law, selfish behavior under the mask of love and also the corruption of personality. The two of the couple, though they possess love, are still two individuals. As an individual, they are unique and in need of private spaces. Some people may say that I earn more than you so I can control you; or I have better family background so I can control you, which makes marriage become beneficial relations, relations between the landlord and farmers, but not couple relations. Some people suspect their spouse and may hire private detective to keep a lookout for him or her. When this idea occurs, love is deserted at once, and the result is often departure, breaking up or discord. These behaviors are not to protect love but to destroy and ruin it.

Of course, when two persons get together, they cannot let things drift. The efficient way of maintaining couple relations is not binding one another but moving them through influence, ability and love. To get married doesn't mean that he or she belongs to your own properties and becomes your servant. The two of the couple are still two individuals with the right of reselecting his or her own life styles. Only reclaim can make him or her realize the conscience; only the ability and self glamour can thicken the relations between your spouse and yourself; and only love can melt all devils. Those who do not receive reclaim and cannot see others' ability clearly are destined to be unhappy, even though they chose another person. And the one who has sacrificed for the other should not feel sorry, just because your spouse is not worthy of your love, you've seen through him or her at last.

Behave oneself respectively. Why do most people often behave properly in courtship and turn to all faults after marriage? This is a question of morality. Any two persons living together should be somewhat self-cultivated and conscientious. If one becomes abnormal after marriage, this is representation of selfishness, deception and immorality. Providing there is still some conscience and morality left in him or her and he or she is still the person in the world, he or she should contribute to love all the time since they get together because of love. It is the fact that some shortcomings will not turn up during courtship, just because "Beauty is in the eyes of the beholders." Thus someone says that you should widen your eyes before marriage and observe things with one eye closed and one eye opened after marriage.

Nearly every man will promise "I will bring you happiness" to the woman while wooing. But did they keep the promise? Many men failed. They ask for many excuses for breaking the promise. Life is transient. If one cannot fulfill the inner promise, his life will be a waste of time in this world.

You only know the object suffers, but never know it's in a bad condition. You only know your own grievance, you are less comfortable than other women and others' husbands are better than yours, but you never know how other women have done in their families. "There is no free lunch in this world." "There is no hate without love, and also no love without hate." No input, no output. Nothing is free to get. You should ask for reason of yourself. Anything needs qualifications, and you need to cultivate yourself. You are not the president, thus you have no reasons not to treat your wife well; you are not as beautiful as the Four Beauties①, thus you have no reason not to treat your husband well.

---

① The Four Beauties of ancient China refer to Xi Shi, Wang Zhaojun, Diao Chan and Yang Yuhuan.

Do not look down upon trifles. Sometimes simple words can ruin the marriage. Basically speaking, there are no large matters but trifles between couples, but it is the ways of solving trifles that bring out different futures of marriage.

Be polite and tolerant to each other. Not everyone can make marriage successful at ease. Marriage is not always perfect all the time. Even Premier Zhou Enlai listed "Principles of Eight Does" with his wife Deng Yingchao. Now let's renew "Principles of Eight Does".

(1) Keep mutual love. As the revolutionary couple, to love each other is the basis.

(2) Keep mutual respect. The ancient people said that the two of the couple should respect each other like treating important guests. Maybe you can follow the rule in the beginning of marriage, but you should pay attention to the following years' life and should not only focus on finding your spouse's faults. You should respect each other, especially in public.

(3) Keep mutual encouragement. No matter you are at work, at school or in life, you should encourage each other and make progress together.

(4) Keep mutual comfort. When you come across unhappy things in life, you should be considerate to each other, but not blame and complain, which will hurt the emotions.

(5) Keep mutual accommodation. There are always quarrels in the family life, and under this condition, you should learn to be courteous and well-mannered, especially when it's not a principled affair.

(6) Keep mutual forgiveness. You should be tolerant to each other. Even if your spouse has faults, you should try to forgive one another.

(7) Keep mutual assistance. The two of the couple should care and love each other in life, work and study.

(8) Keep mutual learning. Everyone has his advantages and disadvantages. We should enlarge the merits and advantages, learn from each other, complement each other and reinforce the emotions.

"Principles of Eight Does" was drawn from several talks between Deng Yingchao and Zhou Enlai. In the very beginning, it's "Four Does", and then "Five Does" and at last "Eight Does". These "Does" became the generalization of the precious experience of their couple life in so many years.

This is the model couple's experience of marriage life and also the representation of their harmonious life. Deng Yingchao felt that Principles Five and

Six were the most difficult in all the principles.

To put it in a simple way, the couple should keep mutual respect, keep mutual accommodation, and keep mutual forgiveness, among which keeping mutual forgiveness is the basis of a harmonious marriage. As the couple, the other five principles are naturally carried out.

Marriage means equality and tolerance on the basis of love. No couples can exist without equality, and no families can stand without tolerance. The basis of maintaining marriage is the mental balance of each other. The premise of a happy marriage is to leave space for each other. No matter husband or wife should not boast of his or her ability though he/she has. The more self-praise you have, the more pressure you'll give the other side. Because as the self-praise accumulates, the contrast between you two becomes more distinct, and this can make your spouse upset. The correct way of adaptation is that when you realize your own ability and achievements, stop self-praise but begin to praise and speak highly of your spouse, thus you can make each other equal, bringing happiness and joy to your spouse. The result is that your spouse can accept your ability and achievements by heart, being sincerely convinced and keeping a joyful relation with you.

Match everydayness with romance. Life is common. Even the president and celebrities' life also contains shopping, cooking, going to work and sleeping. But people are often fickle in affection, that is, love the new and loathe the old. The life which is extremely common or always common equals to dull life. The couple life needs romance, so we should match commonness with romance.

Living together with someone for a long time will cause more frictions. If there is no other focus, then all the attention will be paid to the friction. All human mind has main focus point and goal. Once one focus point and goal disappears, the mind will be transferred to other aspects. To adjust the state of mind, you should try to make new highlights.

Enjoyment is a kind of love which needs reward. The reward may be words or actions. For example, for your lover's contribution, the more you enjoy, the more sense of achievements he or she will feel. This kind of enjoyment is also love for spouses, and it will be perfect to add words and actions of love.

The couple should often consider mutual contributions and merits, and thus you will feel affiliated to the family.

Women are often impulsive. Words of them can be divided into three proportions: one-third is unreasonable; one-third focuses on emotions, and the left one-third emphasizes on rationale. In a word, men should be tolerant of women as there are no other effective methods.

Learn to eliminate conflicts. When you begin one thing, you should make your purpose clear at first and also the principles, measures and steps. Following casual ideas and behaviors can only cause embarrassment. Just like the quarrel between couples, the two should be clear about why and for what they are quarreling. If the result of quarreling cannot fulfill their goal as they expected, why not stop it? But the fact is always that many couples do not know why they quarrel and the result of quarreling. They just stick to quarreling continuously, painfully, and when they come to the result, they refuse it. Unharmonious couples will have three endings: divorce, going on living, and making difficult things for each other. It will be simple if the two decide to break up without long-term torture in the future, bright and straight, with a happy ending. If they are determined to go on living together, they should stop torture and be tolerant, thus an active smile wins back another smile, an active hand wins back getting together, and the two can walk hand in hand, side by side. One warm greeting can win back another courteous word, and the two will be inmate and loving, caring and loving each other, then your affection for each other will be deepened. If the two become hostile to each other and make difficult things for each other, then the result will be that both of them get weakened by the war. It's not rational, as we cannot live on feeling wronged and acting rashly. For the love in the past years, just let it go, as forgiveness can bring friendship and magnanimity can carry fame. Moreover, the new and more wonderful life is waving to you.

Love to the depths is forgiveness. The love in words, the superficial love, and low-level love are not good for strengthening love and maintaining marriage. If you regard love as an excuse, taking one another as your private property, affiliation and even slave and prisoner, you will cause only the corruption of love and break of marriage. Love to the depths is forgiveness.

There are many examples in daily life. For example, one is forbidden to have a word or touch with the opposite sex by his or her spouse; one is always asking his or her spouse's routine in great details, suspecting, doubting, inquiring and secretly tracking, jawing and alarming. It seems that there is no valued focus except for relations between the two sexes in marriage.

Only with tolerance, communications and forgiveness can a couple lead a harmonious and happy life. There were many unfortunate accidents which happened after the quarreling of the couple: the driver may cause traffic accidents; the mechanic may hurt his body, etc. Madame Curie and her husband Mr. Curie were ready for a trip, but Mr. Curie importuned to go to the laboratory to have a look. They had a quarrel which may have caused Mr. Curie's fatal car accident.

5.2 Education on mutual love

As the proverb goes, "Wherever the son goes, the mother will always be missing him, while the son will not worry about his mother even if she had gone for 1,000 miles." This phenomenon has been around since ancient times when people began to have later generations. The root cause of it is the law of inertia. There are two stories below.

Story 1: Mom likes eating fish head

At one time, people did not realize the nutritive value of fish head. When they cooked fish, the mother always chose to eat fish head and selected the left fish meat for her children and husband. Once, the wife, also the mother, was asked by her child, "Mom, why don't you eat fish meat?" She answered, "Darling, because Mom is fond of fish head." When the child grew up, he always left fish head for his mother, for in his ideas Mom had always eaten fish head.

Story 2: Mom has a passion for yam's rind

It was the time when the family was trapped in poverty and lack of grain. When the family sat at table for dinner, the mother always did her housework by side. When the family members finished their dinner, the mother would go to pick up the yam's rinds and ate them up. The kid asked her, "Mom, why eating the rind?" The mother said, "I like eating it." The result is the same as the previous story. When the child grew up, he would leave the yam's rind for his mother.

Why can these occur frequently? The children's ignorance is the result of parents' untruthful teaching and this kind of teaching will hurt children and parents next. "Wherever the son goes, the mother will always be missing him, while the son will not worry about his mother even if she had gone for 1,000 miles." The reason is that parents have been taking care of their children and worried about them till their last days. Years of care becomes the parents' inertia worry that cannot be deserted. If the children begin to worry about their parents when they become sensible, they will form a habitual worry and this kind of worry cannot be deserted in their life. Thus a way of education comes into being: education of mutual love. You should teach your children when they begin to be sensible and let them know something about the society: let them know the fact objectively; let them form a comprehensive understanding of parents' pains and deep love to them; let them know the worries they've brought to parents. Let the children know all the truth, but not the false and processed facts. All of these will be beneficial to the children's growth, making them form the habit of loving and caring for parents, and developing their benevolence, grateful heart, sense of responsibility and the desire to advance.

5.3 Family

Family's a place where all the family members collect both body and mind;
A place where close relatives get together;
A heaven to enjoy the family relationships;
A place named Garden of Eden of love.

Family's a port of the sea after surging waves;
The rest place after long time tired work;
The comfort of mind and spirit;
The base to struggle against life.

It may be a place at everywhere;
It's a forbidden place and can not be taken casually;
It can not be invaded;
It's a sacred palace.

It's the safest area;
And also the most dangerous place;
The place that needs the most adequate protection;
And the place that costs most efforts.

It's the noble place of recounting troubles;
The happy land full of mutual love and encouragement;
And the together ship regardless of situation.

5.4 Parents, did you make it?

The parents and children have different role relations between each other. As parents, you should give the children your love, take care of them and bring them up. As teachers, the parents should give their students (the children) enlightenment education, set good examples with their own conduct, be a paragon of virtue and learning. As the masters, the parents should teach their apprentice (the children) hand by hand and let them complete their apprentice at the appropriate time. As the supervisors, the parents should list the research objects and orientation clearly for the graduates (the children), setting them free to fly in the sky of key knowledge. As the generals, the parents should give the soldiers (the children) military exercises, taking them to reach the goal in front. As the coaches, the parents should make the degree of exercises under control of the athletes (the children): hard but not injured, tired but not harmful, free but not indulgent, controlling but not forbidden. As the connoisseurs, the parents should try to find out the forte, strong points of the fiery steed (the children) and create conditions for developing them. As the model, the children are the followers. You should be a master before being others' teacher. Set a good example for the children and become their model. Parents and children are friends; you should respect each other under the principle of equality and mutual benefit, boasting a confidential exchange of views, baring heart to each other.

Parents, did you make it? Maybe you haven't made it or you haven't totally made it. You have little pains but you want to gain more, you are so greedy. As the kid, he is so little, but he was forced to be No. 1 in studies, to be versatile, fruitful and a genius. Parents, did you make it in your own school life? You've killed the kids' nature, deprived them of freedom, sacrificed their rest time, tramped their wonderful childhood, left the mental distort in their whole lifetime, produced the cause for an unhappy life. Do you love your children? Do you really love them? If so, you ought to give back all that the children should have, adapt to their nature, give them free space for healthy growth.

In 2005, a survey on 17 thousand adults was done in San Diego. According to the survey, the doctors of CDC (center for disease control and prevention) found that the risk of heart disease of those people who had suffered physical abuse, sexual abuse, mental cruelty, family violence, or had family members who are drug addicts, alcoholics in their childhoods, is higher by the rate of 30 percent to 70 percent than the ordinary ones. It clearly shows the importance of the physical and mental health in one's childhood.

Parents should adapt to the children's nature, laws of growth, the detailed conditions, giving the appropriate guidance on the basis of adaptation, but not impelling your will on your children. Children are children; they are individuals with their own thoughts and souls. To treat the children in rude and impulsive way is the simplest way but incompetent and stupid way as well.

Parents should build intimate relations with children when they are young, be their friends heart to heart, for parents are the most important ones in their lives and this intimate relation can be easily formed in their childhood and can last till they grow up. But if this intimate relation was built in their thirteen or fourteen years, it would be late. Once parents become children's friends, the children's problems will turn to be easy. Because you've known each other and can give proper solutions to problems, because you are all getting well with each other, then there are no pressures between you and your children but the motive power.

Parents should spare some time to listen to your children, thinking about questions in your children's shoes, adapting to children's ways of thinking, and understanding your children's views by heart. You should not draw the conclusion too early and impose your own ideas upon your children. You should guide your children how to cherish themselves, what they should do and what they should not do in time. You should increase sensible guidance and decrease detailed guidance in terms of children's growing age.

Let children take part in the process of family decisions, as this is their human right. This can help the children build self-esteem, self-confidence and sense of responsibility.

Encourage your children timely, praise them a lot but reduce times of blame. You should praise, encourage and guide your children to be the kind of person you expect them to be. If it is the condition that children need to blame, you should adopt proper ways and methods, which can protect the children's dignity on one hand and make them to accept criticism on the other. Manage your children in a relaxing but not impulsive way. Just as Lao Zi said, "Ruling a great nation is just like cooking a small delicacy."

In a sense, how many scores the children will get, what kind of person the children will turn to be in the future, whether they are happy or not, all these are in the charge of their parents. Compared with children's education, parents should receive education at first.

5.5 Teach children like controlling water

Water is soft, and water is hard; water is of drops, and water is boundless; water is shapeable, and water is hard to change; water is helpful, and water is harmful; water is clean, and water is dirty; water is under change, and water is stable; water is easy to lead, and water is hard to guide; water is tame, and water is wild; water is still, and water is furious; water is easy to master, and water is hard to control...

Like the water, the children have all the same characteristics. Like the water, the children have infinite plasticity. Control the water according to the water's features and principles, making the best use of the situation. Educate the children according to the children's factual nature and specialties. The water is under your control, and you can use it by ease. The children are under parents' control, and what kind of person they will be is decided by their parents. Like human beings can not control all of the water, parents can not control the whole life of the children.

5.6 Analysis of marriage

5.6.1 Formation of marriage

Affection is the basis of love, while marriage is the result of affection and sublimation of love.

Love is built on the basis of emotions. Maybe you are short in height, venerable in strength, with great distance in ages, but all of these can not constrain the growth of love. However, all love that may not turn to marriage at last. Marriage is built on emotions, interests and proliferation (multiplying). Referring to the traditional marriage, people may think about proliferation (multiplying), focusing on the important part of heredity unconsciously and the part which can improve the species. Thus the deficit (defect) in multiplying always turns to be the obstacle of marriage, despite liking and even love. Family is built on fame, interests and emotions. Family is practical, as it has filtered romance in affection and dispelled the illusions. The wheel of life runs in practical track, it needs fame, interests, emotions and healthy proliferation.

The husband that women always seek is not the smartest one, but the most reliable one; the wife that men often seek is not the most beautiful one, but the favorite one. "Boys will not fall in love with a girl because she is beautiful, but because he likes her so she is beautiful." Marriage, for woman is dependent, and for man is wonder. For women, as they are vulnerable, longing for one to protect her through the lifetime, thus men's kindness, honesty; physique, identity, position, knowledge, wealth, etc. are all the key factors affecting women' choosing their spouse. While men often do not worry much about living, they do not need to depend on others. In actual life, they need those women who are feasts for eyes, courteous and lovely.

Men like the delicacy of women, while women like the fitness of men. Beauty is broad, beauty is profound, and beauty is the supreme grade of objects. Sex, is one aspect of falling for beauty. Females are always prettier than males, from appearance to figure, from skin color to elasticity, from tenderness to kindness.

Finding lovers, seeking for sexual partners are all focusing on novelty, curiosity, beauty and pursuit of pleasure; while looking for a partner in marriage lays great emphasis on tacit agreement, reality, stability, safety, and also living a life.

Mystery and curiosity of sex is the impetus of amativeness, but the pure sexual love is not love. When the two taste the forbidden fruit, the sense of mystery will disappear, and will be accompanied by the loose feelings for each other. The previous intimacy and novelty which maintained and produced the feeling in the very beginning will disappear, let alone cherishing them. The result is that marriage, which was built on the basis of the above changes, would soon turn to commonness or crisis. The so-called "Marriage is the grave of love" is referring to this kind of marriage.

Marriage may not be suitable for everyone, and not all the people may choose to lead a marriage life. In the process of humanized, rationalized, verified, colorful human life, the life styles should be various.

5.6.2 Function of marriage

Sense of accomplishment is, for example, one person falls for you and is lusting after you; or you fall in love with one person and you've got him or her successfully; or you've made accomplishment, your family all celebrate for you and share joy with you, thank you and love you.

Sense of satisfaction is to be harmonious with society and family members, to live a well-off life which all others can lead.

Practicality is to care, help, and encourage each other, being bound to each other, enjoying the sexual love, mutual care, safe harbor, safest roost, and bearing of offspring.

5.6.3 Methods of maintaining good marriage

(1) You should understand that the two after marriage are two relatively independent individuals. Getting married means two individuals get together and form a family. But basically speaking, the two persons are still two individuals with their own thoughts freedom and individual independence relatively. The two can not be united by one same consciousness, one same behavior; thus one can not deprive the other of freedom, but to fully respect the mutual independence. The family members, especially the couples, should give each other enough freedom, give each other space that can freely release the emotions. You should not hijack mutual emotions, blame or demand each other casually, as all of these can cause mental strains, worry and disease at last. A research shows that if the couples suffer from long-term mental strains, then their rate of suffering from heart disease will be higher than other couples who do not. If the heart is sick, then other diseases will come naturally.

The premise of interdependence is to accept each other. To accept one is to accept all aspects of this person. One person may have many undesirable habits that are not easy to give up. Maybe he/she is not willing to own these habits, and this is not his/her faults to keep these habits, but they are the reasons of natural inherit by generations. As his or her partner in marriage, you have only one choice, that is, to accept it. Since you two get together for the reason of love, no matter merits or shortcomings, you should accept them all. For everyone has his own merits and shortcomings, and the person whom you like is a live human being, you can not only accept his/her merits and hate the shortcomings. The combination of the couple is the combination of emotions.

To maintain this kind of emotions, you should cooperate with each other, forgive each other, understand each other, look after each other and depend on each other.

Marriage is the life style that needs to bear the offspring under the social systems and meet the sexual need, and the combination of mutual interdependence, care and being win-win. Where the care lies, the human exists. With the improvement of people's material living standard, marriage lays more emphasis on mental life. The basic elements of maintaining a good marriage is that though you are in high position, the politeness goes first. At the same time, you should make each other satisfied in the aspects where he/she cares most and you should give up selfishness, e.g. being critical, extortionate, self-centered, etc. And all of these are the basic elements of a good marriage.

(2) Marriage is a glass of water. At the beginning of marriage, the water is clean and clear to see through. With the time going on, water can not always be clean; dust will fall into it and it will get dusted and mixed. The mixture will fall down to the bottom of the water, but the water can still be drinkable. If this discipline can not be followed, then the water can not be clean from then on. A little dust can not be tolerated, because it must be followed by a big quarrel, and then the unclean water can not be safe to drink. Have you ever mixed the water?

Marriage is a glass of water. At the beginning of marriage, water tastes sweet. With the time going on, different tastes of life will be added to it. If the ingredient is of delicious smell, then the water is sweeter to taste. But if the ingredient is of boring smell, then the water will not be able to drink. What taste of ingredient do you add to your marriage?

Marriage is a glass of water, the mixture of which can be filtered; the unpleasant smell can be eliminated. And the key method comes from honesty and purity.

(3) Marriage needs to be managed. What is management? Management is that your input can get output; you make efforts and then you gains; you have promising and you get your goal; you are united and then you will be prosperous; you are live and vigorous and then you can develop; you have benefit and then you can survive. The marriage is that one plus one always exceeds two ($1+1>2$). Marriage needs mutual non-stop efforts to preserve, including self development, selfless service, caring one another, communicating with others in time, tolerance, seeking common ground while shelving

differences, maintaining fully inclusive and equitable, concerning inner and outer part together. Attraction before marriage comes from sexual love, while attraction after marriage mainly depends on development of emotions. People always attach more importance to one word "honesty" on the critical relations, for example, doing business, employing workers, etc. People believe one word in the communication, that is "honesty", and also on marriage, the same word "honesty" weights a lot.

What is marriage? Marriage is a career that can not be easily fulfilled. Marriage is the responsibility that is not kid stuff. Marriage is giving, not taking from others. Marriage is experiencing, the experiencing and sublimation of personality, tolerance, degree of forgiveness. Marriage is the discard of bad things. Marriage is friction, the friction of two individuals, two personalities, two interests, two needs, and so on. The friction is ubiquitous war between the two paradoxical individuals, who can never be combined into one person but still keep together all the time. Marriage is to adapt to each other on the basis of following the principles, but not seeking to prevail over others, not failing in cooperating with others. Adapting to the elders is filial piety; adapting to the spouse is love; adapting to the children is the best method of education. Marriage is a test, the whole-life test on mutual merits, love, loyalty, diligence, cultivation, will, etc.

Marriage needs balance. The family disharmony often occurs under the condition that one of the two gave too much without receiving any returns, mainly referring to understanding, thanks and forgiveness, so he or she would lose balance in mind at last. The solution to this question is to make self give less and the other give more so as to understand and forgive each other.

The demand of happiness in marriage grows with the time. Modern females enjoy more freedom than before, becoming more individual in economics and depending less on males; thus marriage will be less stable. A good marriage is made of half happiness and half sadness, while the bad marriage is that sadness surpasses happiness or marriage becomes the grave of happiness. Marriage can let people enjoy the happiness of family union and also the sadness of family union at the same time. Therefore, we must take an objective attitude towards it.

Be clear about the goal of marriage. The key elements of doing one thing are: goal(purpose), the relevant persons, objects and actions. When you set the goal of living with another person happily and harmoniously, the actions should be beneficial for achieving your goal, and vice versa. The saddest thing is that you do not know what goal you should act for, so you toss about all day long, and finally you can not put away all the things and get regretted.

(4) Nine key words for a good marriage. Love is the basis of marriage, and as long as you have built up this basis and fulfilled the nine words, you can keep a long-term harmonious and happy marriage. The nine words are: kindness, harmony, leniency, tolerance, comity, forgiveness, respect, care and praise. Kindness is being easy-going and kind, and harmony is being harmonious and adapted to society. Leniency is being lenient. Tolerance is being tolerable, as tolerance can patch up a quarrel and reconcile the concerned parties. Comity is being courteous; being courteous then you will become lenient. Forgiveness is to forgive, and only forgiveness and understanding can solve the problem successfully. Respect is to respect each other all the time and care is to take care of each other and praise is to use more words to praise each other.

5.6.4 Circulation of marriage

The cyclic development of all things is from lower level to higher level. The seasons of a year circles one by one, and new meanings are added every year. "After generations people would still come and go", and the new generation will be more successful than the last generation. With the turn of dynasties, the social civilization is promoted.

The forms of marriage are no exception of the circle of development. Matrilineal Period—Paternal Period—Polygyny—Promiscuous mating (Actually the common obligation lies in the children, but the children may not be one's own)—No getting married (Only bearing children, with features of Matrilineal Period, at the time when males are looked down upon and the sons can desert their fathers. The disabled, in case that the next generation may get disabled and suffer from it, accepting the direction of social public opinion, give up bearing children)—Polygyny (The males establish their authority, fame, status, focusing on their economic benefits and inherit issues) —Promiscuous mating (The nature of human causes this phenomenon in an open society)—No getting married.

There are two reasons that marriage can develop in this way. The internal cause is the nature of human, while the external cause is the social economic surroundings. Look at the following news report!

Reported by British State Statistics Bureau, recently the rate of getting married in Britain was way down. And if the present condition continues, the married couples will become minorities in 24 years in Britain.

For ten years, the live-in partner amounts to 2,300,000 pairs, raised by 22 percent, while the married couples reduced by 4 percent, which amount to 12,100,000 pairs. The number of the single mother has increased by 8 percent, reaching to the number of 2,600,000.

In the present family in London, family of single mother has taken up 22 percent, which is higher than the other areas in the whole country.

The development of marriage condition in Britain is the forerunner of other countries in the world.

5.7 On Love

5.7.1 Definition and origin of love

Love is the furious passion between two sexes. Human is the combination of spirit and body, and love is nothing but the most exciting part of emotion, the best feeling and the best combination of mutual spirits and bodies. Love is the combination of emotion and sexual love. The pure combination of bodies and pure combination of spirits are not love. Beauty plus tacit agreement equals to love. And beauty contains two aspects: physical beauty and spiritual beauty. Tacit agreement is an inspiration. "Lovers' hearts are closely linked." Sex is the body of love, while spirit is the soul of love. The soul conducts the body. For example, when the man and the woman fall in love with each other, they would feel some chemicals between each other when they touch each other by hands. But after marriage, the persons are the same as before, the chemicals are gone. Even when they have a hug, they can not find the feeling they have ever had before. It is the spirit that makes the difference.

Love is caring about one another, keeping one another from any hurt, grievances, and sadness. Love is devotion, being willing to do any (legal) things for each other, even sacrificing one's life. Love is tolerance, never fussing about each other. Love is undertaking, being willing to undertake the duties. Love is force, not hurting someone by actions with love in heart, not depriving others' freedom. Love needs basic training, practice and management, but it is not anyone's casual action.

A pleasant conversation amidst flowers and in the moonlight, the dance in the musical hall, the ramble on the avenue, the life and death lingering, the warm hug, the attractive kiss may not be love, but love can be expressed in the above ways. Beauty is the most graceful and sacred thing in the cosmos. Love is the most graceful, sacred and beautiful emotion. Love is not the rope, not the prison, not the ordinary thing. By contrast, love is sunshine, happiness, tolerance and infinite grace. Exclusiveness, selfishness, fighting with rival in love, hijacking spouse are not the patent of love, but embodiment of possessiveness. As all of those may be done by animals, but those animate behaviors are not equal to love. Love is the lofty emotion between humans, and the supreme realm of it is to love each other, hoping each other to live a better life both in emotions and life. And this is true love.

Love starts from two aspects: body and spirit. The pure wonderful sexual love can only satisfy the brutal desire, while the pure intelligent thinking can only satisfy the spiritual desire. Love usually originates from sexual love, which weighs a lot in the beginning. With the development of the love affair, the proportion of sexual love reduces and the emotions become calm. At this time, the proportion of affection will rise. Affection and sexual love will promote each other. Sexual love will be strengthened with the promotion of the facial elements, like appearance, status, forces, etc. Affection will be promoted with the promotion of inner factors, like behaviors, ability and inner quality. Along with the increase of age, affection turns to be love and this is interdependent love, tolerated love, mutually known love and one-heart love between two people.

5.7.2 Features of love

Love is like water. Love is eternal, love is transient; love is stable, love is changeable; love is individual, love is various; love is single-minded, love is generative; love is indolent, love is flexible; love is selfish, love is selfless; love is rigid, love is tolerable; love is conservative, love is romantic; love is common, love is sacred. Love needs vigor, development and sublimation. As people are various and changeable, and the environment is changing, the emotions, which depends on people, are not stable, so love should be grasped dynamically. The result of love is in the charge of each other. Love is wonderful, but when the two feel not happy as before, love will be lost. Love originates from the combination of spirits and bodies as well, but ends in the combination of spirits and bodies as well. Just as one saying goes, "Either success or failure boils down to the same person." [1] Since the fusion of bodies and spirits can produce love, then there will be more suitable combination of spirits and bodies. Love will be transferred, and if there are many possibilities of combination, then love will be more. The common feeling can turn to be love and love can also turn to be common feelings. Love is one emotion that can have innumerable twists and turns.

Love is a double-edged sword. Love is blind, love is dumb, and love is a silly guy. Love does not care the present, love ignore the possible consequence, and love is a bad guy. Love is not afraid of poverty, brutal suppression, and love is a stupefied guy. Love is besotted, love is obsessed, and love is a maze. Love is tacit understanding, love is electric wave, and love is a deep card. Love can relieve the pain, love can cure the disease, and love is the most efficient doctor. Love is magic, love is divine water, and love is omnipotent.

Love is never perfect. Here is a story: A lady vowed to herself to marry to a perfect man. She went here and there to seek the perfect man. Time went on, and at the year when she turned to seventies, she was still single and led a hard life. One day, someone asked her, "Have you ever found a perfect man in so many years?" "Yes, I did have, but he also seeks a perfect lady!" the lady answered sadly. There exists no perfect person in this world, and no perfect feeling. Even though it exists, it will change with the imperfect parts of people. Shortcomings are like the mole on woman's face, someone hates it, but someone takes it as a beauty mark.

People often ask this question: does true love or eternal love exist in the world, especially in the crisis of marriage? With the independence of living ability and development of individuation and humanization, this problem will be played down, and the eternal love will be inconceivable for the young people in future. But, certainly, true love and eternal love do exist, that is to produce sacred love on the basis of the continuous combination and sublimation of mutual bodies and spirits.

---

① "Either success or failure boils down to the same person." This is the classic description of Xiao He, the folk hero of the Western Han Dynasty. Xiao He recommended Han Xin to be the major general; Han Xin was killed due to Xiao He's plotting.

5.7.3 Methods for maintaining love

The only method of keeping love is to keep the good state of fusion and combination of mutual bodies and spirits. One can not be interested in a person who is careless about his appearance and stays dirty in a long term. One can not maintain love on the person who has extreme bad ideas, being quite different from you in all aspects. In a word, love is mutual, and so is love maintaining; love is changing, the two who combine with each other also need to develop according to the specific situations to satisfy the need of love. Good look can induce one person at transient time, but that does not last long. The internal grace can weigh ten times the appearance. Maintaining love needs improving the level of affection.

Loving one is not to change him or her, but to adapt oneself to him or her thoroughly. For at the very beginning you fall in love with him or her, you feel addicted to him or her and let him or her do what he or she likes to do, but not to change him or her. Nevertheless, nobody will love you in order to change himself. The ideas of changing is the beginning of being self-centered, aiming for one's own purposes. Thus, love is adaptation, and following with each other. Following is love, and love improves on the basis of following to reach the further harmony.

Love is not wild and rude but gentle. It needs care. Love is generous. If you treat one another honestly, and can undergo the tests, then you can gain love. "To love one is not to break his wings of flying, but let him fly higher. Giving your beloved one freedom is to give yourself freedom and happiness. Only knowing to love without giving love freedom can not be called love, but a kind of burden and line. This can not be embodiment of love, but discard and forgetting of it."

5.7.4 Even among brothers, accounts should be settled without ambiguity

There are two important things between brothers: one is to support the old, the other is the struggle of economic interests.

Reckoning is to be clear in the facial and obvious aspects without ambiguities. In the outer occasions, when all the brothers are in, they should undertake the responsibility, pressure and duties and make clear about the above aspects to show the equality: they are all good sons. For example, in the aspect of supporting the old, the brothers should share the burden together. Then in other aspects, the brothers can still help each other.

The things between full brothers should be made clear and rules should be set at the very beginning; otherwise, troubles, misunderstandings and even contradictions would occur.

5.7.5 The softer the emotions are, the more powerful they will be

In emotions, the weak and the poor can own more power.

Delicacy makes others cherish you, while stiff makes others hate you. The wife should behave delicate, easy-going and gentle between the two, as virtue is the most powerful weapon to win the man's heart. Harshness and shrew can not win the man but can be defeated by the man. There is a popular saying going like that "The dear delicate wife is the most adorable wife in men's hearts." However, if the husband is rude, the wife can not behave delicate. The masculinity is not harshness in emotions, and the women do need gentleness as well. The women's gentleness can conquer men, and men can also conquer women with their consideration and gentleness.

Heroes do not fear death in the fighting field, but they are often defeated by beauties. There is part of sexual lure in this aspect, which lays great emphasis on the conquering power in the delicacy. Women, especially young girls, should hold tightly to the character of gentleness, and then they will grasp the most powerful weapon of attracting others' care and consideration.

Tears are soft but often powerful.

Sincere words and earnest wishes are soft, but the effect they carry is more powerful.

Compared with the rigid criticizing in the public areas, criticizing in easy-going words in the back will bring about the most grateful emotions, and this can not be realized by harsh attitude.

The beggars are soft, and they can even make the masculine man merciful.

Friends, especially the opposite sex friends with low origin and poor family background, can bring forth others' sympathy and can also make the relationship more intimate at once.

Those whose life is full of misfortunes are the weak, and they can touch people easily. It is true that many frauds are familiar with the special character of people and cheat people at times.

Politeness and courtesy are soft, compared with the ignorance and rudeness, and they are much better accepted by people.

Kindness and softy are the core of emotions and the accumulation of wealth.

Emotions like affection, friendship, patriotic feeling, are motive for everything. The output of emotion often need the whole life, but the nature of emotion is soft.

Human beings are the emotional animals, full of passions. And good emotions are precious and soft. The softer the emotions are, the more powerful they will be.

5.8 How to be filial

Filial piety is the first human nature. The old saying goes like that in all the virtues the filial piety lists the first. Our parents have paid great efforts to give us the bodies and lives, teach us and bring us up. Without our parents, we could not have our lives today; thus supporting our parents is unconditional. Parents' love is like the sea, and we can not repay them the whole life. As the proverb goes, "Being grateful is a small virtue, but ingratitude is the worst of vices." Those who can not do filial deity can not be said to have human nature. Often there are four types of persons that can not be made friends with: the one who is disloyal to the country; the one who is not filial to his parents; the one who doesn't love his spouse and the one who does not remain faithful to friends. We can not count on the one who is not filial to his parents to be kind to others.

Then how to do filial piety? The conclusions made here are: realizing parents' dreams, adapting to their minds, and satisfying their needs. We can make them more concise into five words: independence, respect, adaptation, supporting, and suggesting.

5.8.1 Realizing their dreams

The parents often have two wishes: the first is wishing their children to live a better life and the second is that they think that their ideas are more helpful for the children. As the children, we should fulfill the following things:

Manage yourself well and realize yourself. Try to do well with your own studies, jobs, making your parents free from worrying too much. Try to be independent, making your parents free from giving too much. Try to take good care of yourself, refuse to be contaminated by evil influences and keep yourself out of troubles, making your parents free from worry. This principle represents the first word: independence.

Support the old and comply with their wishes. First, the old often lay many wishes on the young. Neither should we obey them all nor should we do not listen to them at all; it is unsubstantial, not correct. We should respect our parents' suggestion, trying to take their advice into consideration, and to give consideration to the whole situation. Second, everyone is longing for self-realization. The old spend the whole lifetime realizing themselves, they are in need of the children' supporting and carrying on their cause. Making their cause forward is the half meaning of their life. And this principle represents the second word: respect.

5.8.2 Adapting to their minds

Adapting is not an easy thing. It needs skills and efforts. There are two aspects in adapting to parents.

Adapting is the basic principle of filial deity. As the saying goes, "100% filial piety can not be equal to 1% adaptation." The filial piety lists adaptation in the first place. To be filial is to adapt, and adaptation to the parents is the real filial piety. Do not object to the old with rigid manners. If you do not follow the ideas and manners of the elderly, though you have done thousands of good deeds for them, your filial piety will be greatly abated. There is a saying goes like that obedience is better than politeness. The more politeness will turn to nothing under disobedience and not adapting. When we communicate with others, we often carry out this principle of adapting, and then why not treat our parents in the same manner? This principle represents the third word: adaptation.

Adapting should avoid emphasizing one at the expense of another. Adapting is to adapt to the logos and the principles; adapting is lofty but not parochial. Accommodating oneself to parents blindly only does harm to parents, which will break the nature of adapting. "To err is human." Parents sometimes need remind and advice. And this principle represents the fifth word: suggesting.

5.8.3 Fulfilling their demands

The needs of parents lie in three aspects: material, mental and physical.

(1) Meet their material needs, ensuring their well-off life. To bring up children is to make preperation for one's old age. The parents get old, but if they have no savings, if they demand something in life, the children should offer all for them and this is the primary filial piety for supporting them.

(2) Meet their mental needs, ensuring their pleasant life. If supporting the old only means giving a little money, then why did not they choose to buy blue chip? The present senior citizens are lack of mental comfort, mental support at the largest part, and the reason to a large extent is that the material condition is satisfied, but the mental aspect is inadequate. Human beings are emotional animal; they are in need of emotional patronage. Especially in the modern wealthy life, mental need becomes more important. The mental needs contain mental life and mental satisfaction.

Mental life contains spiritual and sense organ pleasure, for example, listening to music, watching plays, reading, painting, calligraphy, dancing, etc. As the children, we should prepare all the activities for them.

Mental satisfaction contains the maintenance of self-value and respect, for example, making the old feel they are useful and keeping their authority in the family, etc.

Here is an example: It's the son's treat. After dinner, the son asked his mother, "Mom, could you do me a favor washing the dishes?" The mother answered, "OK, put them there." The mother said pleasantly, "I always wash the dishes cleaner than you all do." But the friends asked, "Why bother your mother to wash dishes, we can help doing that." The son said, "Giving the proper work to the old can not only exercise them, but make them feel useful, and then the old will feel pleasant." Everyone was suddenly enlightened.

Somebody does not think of this aspect of question, they want the old to lead a well-off life, and let them do nothing. Then the result is that the old begin to feel useless; they begin to have strong sense of being deserted occurred, and then the sense of solitude and loss. You should put yourself in their place and care them more. You should let the one who you want to care more feel they are supported in the thoughts. The only way of good care is to put you into their position. If the care is not given correctly, the result will be a regrettful one.

In addition, you should tell your cause, decision, conditions of your work and life to your parents, making them feel your trust on them, avoiding the feeling of seeing the growth of the children on one side, sensing the disappointment on another.

(3) Keep the old healthy, ensuring a long life. The body depends much on the nutritive materials. When the man gets older, all his organs suffer from atrophy and function degeneration, and the integrated balance becomes hard to keep. And then it is time to do health care, studying health care and doing comprehensive care. Have health examination regularly, preventing the body from incurable illness.

All in all, we should be considerate to our parents in emotions, support them in life carefully, and try to follow their ideas in affairs-handling. Among the five-word principles of independence, respect, adaptation, supporting and suggesting, adaptation is the basic principle which means adapting to parents' will, needs and wishes.

In the Buddhism, there are three levels for filial piety. The first level is to support the parents under principles, supporting the parents in a proper way, being free of material shortage. The second level is to bring honor to your family and clan, be honest, be successful in causes. Filial piety on the third level is the supreme level, which is to lead your parents to have good morals, to be merciful with regions, then they can not only set themselves in society the whole life but can have a fine place to go after passing away.

There is a saying form Mencius: "There are three major offenses against filial piety; bearing no descendant is by far the most unforgivable. Shun, the monarch in ancient China was married without informing his parents because he was afraid to have no offspring; therefore, men of noble character considered that his doing so was the same as he had informed them." In *Thirteen Classics of Confucianism*, book of etiquette and ceremonial, there is note beneath "bearing no descendant is by far the most unforgivable": "Refer to politeness, there are three kinds of impieties, listening to parents blindly and making them be trapped in injustice is the first kind of impiety; living with old parents under straitened situation without becoming officials to get salary is the second kind of impiety; not getting married and bear no offspring to inherit, the third kind of impiety. And among the three, bearing no offspring is the worst offense against filial piety."

5.8.4 My view on quality education

Life is transient and long, and there are many turns in life: turns between day and night, turns among four seasons of the year; turns between fortunes and misfortunes, turns between honors and dishonors, turns between gains and loss, turns of students at different periods, turns of different jobs, and also possible turns of different marriages, etc. The things that are done several times are numerous, and you should make up the lessons which you lack. You always have the time to make up, even if you've paid great price.

What is human? Marx said, "Human is the sum of social relations." Living in society, human is a social person first. When people refer to one person, they think of his social position, his social relations first, but not his flesh body. As a social person, the principal question he faces is how to live in society. And only after living can he mention the development, achievements in causes, and contribution to society. The fittest man can survive. As the sum of social relations, people lives in social relations, which are his dependent basis and surroundings of survival. Therefore, he should adapt to the social relations first, establish the social relations well (relations among people), and then he manages to survive.

The principal task of education is to enlighten the life, to help people to be independent, to grow up healthily, to make the idle successful and make him contribute to society.

Man needs a lot of knowledge in his own life. In terms of the degree of importance, there is a center and four surrounded circles, and the degree of importance goes down from the center to outer circles gradually.

The central education is education on the quality and morality of the person, and the detailed education is education on kindness, harmony and forgiveness. Kindness is to be kind to oneself and treat others kindly; harmony is mild-mannered, getting well with others and being harmonious; forgiveness is being forgivable to others, being tolerable. The personality and glamour, the strong will and good character of one person are the decisive factors in his success in society.

Education in the first circle is education on adaptation to society and social communications. This circle is to cultivate the basic skills of man to stand in the society. There is a popular saying goes like that 30 percent of success depends on knowledge and 70 percent of success counts on relations. Maybe this statement is a little bit exaggerated, but here the relations not only refer to the negative meaning of relations in a narrow sense, but also refer to the ability to deal with complex social relations. Success depends on organizations and society, as organizations and society are all human communities, which are full of interpersonal relations. If one fights alone by himself in society, he can only meet failures.

Pay more attention to the education and practice of mental ability to adapt to society, the ability to bear, the ability to differentiate truth from false things; focus more on the education of practical application, practical operation and practical combat exercises; provide the children with grading education according to their physical features and habitual nature. Do not provide distorted education, such as taking the children as adults and smart children as child prodigy, separating more mature mind children from the other "normal" children, depriving them of rights of enjoying innocent life and entertainment. Do not repress, bind and constrain children; do not ruin their natures; and do not praise and boast children blindly, or blow and satirize children that make them raise mental obstacles of fearing failure and world-weariness.

There are many ways of experiencing in social relations. Firstly, you can make full use of the human resources in the campus, adopting various means to make the students have a comprehensive understanding of all the students' families and individual backgrounds, appraise and encourage each other to make progress. You can also extend the range to two classes or even the whole campus, creating more chances for the students to communicate with people. Secondly, you can create chances for children to pay on-site visit and get to know the situation of all walks of life.

The three "child prodigy" in Class for the Gifted Young, Grade 1978 at University of Science and Technology of China had obstacles in communications with others, which is the reflection of the importance of the education in the first circle.

One of them was unwilling to do his professional physics studies, suffering from the pressure brought by the name of "child prodigy". He evaded and gave up three successive examinations for Postgraduates after 1982. He failed in TOEFL three times in 1989, 1990 and 1991 and later he learned to do *qigong* (breathing exercises), became a vegan and kept away from the social life. In 2002, he went to Wutai Mountains and became a monk, completely cutting down his ties with society.

The second one entered the university when he was 11 years old, and the teacher began to worry about his curriculums of interpersonal relations and mental health since his enrollment. At that time, the whole class was lagging behind in the two curriculums, and his problem was rather serious. He followed Guangzhao Chou, the Vice-principal in Chinese Academy of Sciences to undertake PhD in his year of eighteen. Because of his unharmonious relations with the supervisor, he did not get the PhD degree successfully. Later he went to American Princeton University and became a student of a famous professor Philip Warren Anderson; there he went on undertaking the PhD. It is said that he has ever threatened Anderson with a handgun or a kitchen knife, and then the Vice-principal of Chinese Science and Technology University decided to call him back to China. Till now, this thirty-nine-old associate professor seems nearly unfamiliar with ways of communications with others.

The third one shares unbelievable similarity with the second wonder one. Both of them have stayed in Princeton, majoring in theoretical physics, and being nervous on the relations with their supervisors. The only difference lies in that this person made a complete escape. Actually, he lives in He Fei (the provincial capital of An Hui province), making his living with his mother, being completely isolated from the world.

Nowadays, it is commonoccurrence that some students killed their parents because of parents' criticism; some students killed their teachers due to the teachers' criticism, and some students killed their classmates for jealousy.

Education in the second circle is the education on philosophy, politics and current events. Teach the children how to recognize the universe, the world and the things; teach them how to look upon society, the relations between persons; teach them to get to know the domestic affairs and world affairs; put them in the atmosphere of society; help them build up the correct outlook of life and establish lofty ideas; develop their ability of recognizing and analyzing problems.

Education in the third circle is cducation on professional knowledge. The professional knowledge is the specialized subject which is set by the person himself, and with the help the family members and teachers, etc. It is the first choice of future career, and also the first vocational skill to adapt oneself to society. The education on the professional knowledge should be carried on according to one's merits, and it should be divided into fast lane and slow lane. The result will be that in the fast lane the talented children will get what they want, and in the slow lane the slow children will not lag behind. The present educating method is to tell the children that they can not become achieved people if they do not bear the sorrows in life by giving them pressure firstly, which will press them all the time on the way upward to his social position. This education should be changed in this way: parents should take the children to pay the on-site visit to all walks of life, all kinds of jobs, and various work conditions, making them get familiar with the vocational skills needed in the jobs when they are young, or teach them by means of videos. From these practice, the children can choose their developing ways according to their hobbies. Carry the practical operation with practical questions, and make the children learn with zero pressure. The time of being children, students, and young generation are the periods to lay foundation for life; thus any exercises and jobs will be beneficial, as in these times all of these can improve the learning interests, positivity and efficiency greatly.

Education in the fourth circle is education on vocational interesting subjects. Education in this circle is hobby, the supplement and improvement of the second and third professional subjects, the research object in wider fields, the cradle of comprehensive talents.

I think education in this way is the real quality education.

Chapter 6 Enlightenment

From the earliest times, the world gives a lot of enlightements to people. And today I' ve got some new enlightenment. Read this chapter and your life will be easier.

6.1 Collocation is an eternal principle

The whole universe and all the concrete and abstract things in it exist and develop by collocating with their respective antitheses. Relevant sides of a thing supplement each other, while opposite sides of the same icon are interchangeable, for and against each other. So, collocation accords with the development laws of things whenever we're doing or thinking about things. Collocation has super efficacy; collocation makes miracles.

6.1.1 All things need to be collocated

Both the internal factors of a thing and its external linkages need to be collocated.

Things are achieved through collocation, like formulas and secret recipes, scientific diets, scientific intercropping of crops, mechanics, aesthetics and tectonics of any construction engineering, schools' training courses for people of different talents, close cooperation among all branches of arms in military actions, productive technologies and assembly lines in industry, and the business management's rational combination of products, price, market, promotion measures, after services, etc.

6.1.2 Collocation mainly refers to a thing's collocating with its antithesis

Things exist and develop along with their antitheses; a thing's two opposite sides always collocate with each other.

Finitude collocates with infinitude: the concrete are finite while the universal are infinite.

Extinction collocates with immortalization: concrete things will be extinct while substance is immortal.

Motionlessness collocates with motion: motionlessness and motion are relatively defined.

Birth collocates with death: all things inside an object continuously experience birth-and-death. And in the end the object will die, which will lead to the birth of another object.

Food collocation: human beings need various kinds of nutrition so that our food can not be simplex. It must realize the collocation of meat and vegetables; refined grains and roughage; good and bad food; palatable and unpalatable food. Meanwhile, it has to correspond with nutrition sciences, one's physical conditions and work. Only in that way can food not be wasted and can a healthy body be achieved.

Work collocation: different projects in our work, professional and amateur work, set work and hobbies, as well as principal work and avocation have to be collocated. Only then can our work be animated and improved, and can we be happy.

Education collocation: when educating children, parents should collocate management with dotage, care with freedom, assistance with independence, success with lessons, goodness with badness, school activities with social ones. If so, parents can effectively educate and train their children. When children need help, parents should be present; when children don't need help, parents should let them go. Anyway, while parents care children all along, they should set children free as far as possible.

Love and marriage: a marital life should conclude wateriness and romantic, accommodation and correction, tolerance and fight, indulgence and supervision, care and independence, enjoyment and struggle, together with pamper and suppress. That's how a couple to live harmoniously and happily.

Collocation between men and women. There is an old saying: "it won't be tiring when men and women work together." Men and women should coexist and cooperate within a leading group, or when engaging in social activities or an artistic performance. Only in this way can men and women be both taken into consideration and can our society be balanced and stable.

Collocation between good times and challenging times: first, when in good times, we should always remember the possibility of challenges; however, when in adversity, we must possess the hope of prosperity. Second, adaptation and surrender are not our only answers when we are dealing with people or affairs; we have to adhere to principles and consider our own opinions.

Collocation between brainwork and physical power: brainwork and physical power can promote and progress together. So, people who are engaged in mental labor should exercise to build up their bodies; those who work with their hands should use their brains frequently. If a mental worker don't take exercise, he will never be energetic or active; and if the manual worker forgets to use his mind, he is bound to stumble for the increasing labor output. When succeeding in improving both his intelligence and body contemporarily, one can be a whole achiever.

Collocation between favors and humiliations: one will be spoiled if he is unceasingly favored; his psychology will be twisted if he is constantly humiliated. A man has to be both favored and humiliated to grow up healthily.

Collocation between work and rest: overwork leads to illness, and excessive leisure causes a state of desolation.

Collocation between "give" and "take": If one wants to take, one must give first.

Collocation between glory and frustration: when you are in your glory, frustration has come before your eyes. When you are in frustration, glory becomes close to you. So, you have to keep frustration in mind when having your moment, and vice versa.

Collocation between tension and easiness: one who is used to living high on the hogs and eating his head off will definitely have problems of spiritual barren, loneliness, emptiness, and physical disturbance. Yet, overstress can bring you nothing but heavy loads which can cause a total physical and mental eclipse.

Collocation between being busy and leisure: it is always right to snatch a little leisure from a busy life to alternate tension with relaxation.

Collocation between escape and counterattack: sometimes escape may put you in a passive position of having to receive blows. And if you dare to summon your courage to fight back, things will probably turn around for you.

Collocation between the big and the small: big things are not greatly fearful because they are all composed of small things and they are controllable regardless of their size. However, we can never despise small things for we see big things through small ones and a solid dyke can collapse because of an ant hole in it.① All in all, people have to coordinate the big with the small to serve their purpose.

Distance collocation: whether being far or near, things can both be safe. In some occasions, being near can keep us away from danger or hurt, while in other occasions being far is more favorable. A man who always keeps a distance from other cars when practicing his driving skills will look awkward. Whereas, if he dare have his car close to others' on a certain basis, his fear will soon disappear and he himself will pick up driving skills finally. Keeping the king's company is tantamount to living with a tiger②, but meanwhile one can become the person more needed if he is more closely connected to leaders.

Collocation between inner life and external life: inner life and external life are two relative terms. If a man lacks thoughts, wisdom, and reflections in his inner life, he is basically subject to a loser.

Everything has its antithesis to collocate each other. Other examples of collocation include: heavy load and light packs, lofty tine and low key, obedience and rebel, crankiness and convergence, immediate interests and long term returns, *Yin* and *Yang* (the two opposing principles in nature, the former means feminine and negative, and the latter masculine and positive), significance and subordination, worry and relief, fast and slow, old and new, first and last, back and forth, before and after, hero and coward, tall and short, long and short, stand and sit, bitter and sweet, cold and warm, smart and stupid, rise and fall, implicit and obscure, shy and outgoing, strong and weak, abstraction and specification, rich and poor, expensive and cheap, diligent and lazy, good and bad, etc.

In short, everything corresponds with its opposite. A man should always bear them two in mind at the same time. Collocation is invincible.

Here I present you two instances for enlightenment.

Example 1: Joy collocating with sorrow

In the East Jin dynasty (317—420) there was a well-known prime minister named Xie An③. One day, he was playing chess with another man when a

---

① Quoted from a classical Chinese literature book *Han Feizi*, which means: "A small leak will sink a great ship."

② Quoted from a classical Chinese book written by someone in the Qing Dynasty.

③ Xie An(320—385), the prime minister of the Eastern Jin Dynasty.

battlefield report was delivered to him. It was about his son's leading the army to fight in the warfront in Feishui County. After reading it, Xie An continued to play chess without saying a word. His chess partner, seeing him in still, misinterpreted that things went in a bad way. So he made cautious inquires about that battlefield report nervously. Beyond his expectations, Xie An replied: "they gained the battle", being calm and composed at ease. When fighting in battlefield, some may survive while some may die. In war, sometimes there is victory and sometimes loss. Xie An was not complacent on the momentary victory but laid importance in helping his son avoid possible failures.

Example 2: Worries of the Fisher King

There is a fisherman who is known as "Fisher King" due to his first-class fishing techniques. However, the "Fisher King" becomes very distressed at his old age because his three sons' fishing skills are too mediocre.

So he regularly told his troubles to others: "I really can't understand this. My fishing techniques are so good, why my sons' so poor? Ever since they could remember things, I've taught them my fishing techniques from the most basic things. I told them how to weave nets to catch fish in the easiest way, how to row a boat without disturbing the fish, and how to cast a net in order to set traps for fish with the least efforts ... And when they grew up, I also taught them how to recognize tides, identified fishing season ... I unreservedly taught them everything I have summarized through years. But it turned out that their fishing skills are worse than sons of other fishermen whose techniques were worse than mine."

A passerby heard his story and said: "Have you instructed them through teach-by-doing all the time?"

"Yes, in order to make them have the first-class fishing techniques, I teach them very carefully and with great patience."

"Have your children followed you all through?"

"Yes, to help them to avoid detours, I have always kept them following me to study."

Then the passerby said: "then, your error is quite clear. You only taught them techniques instead of lessons. Lessons are just like experiences. It is the source where people gain skills."

Have you ever made the same mistake as the Fisher King's when educating your children? No matter how strong your skills are, what belongs to you belongs to you only. If you really want your children to learn some skills, you can't always take the teach-by-doing method. Children have to grow up with the company of failures, in order to accumulate experiences and skills and gain the essence of success.

6.2 Antitheses are complimentary and interchangeable

Things are made up by antitheses and everything can not live without antitheses. Antitheses depend on each other and are interchangeable. A thing always develops towards its antithesis, which is decided by the law of "things are made up by antitheses".

6.2.1 Antitheses are complimentary with and reciprocally transformed to each other

Blessings and curses: "good fortune lies within bad; bad fortune lurks within good."[①] Blessings and curses take each other as the basis. Many people whose childhood were very unfortunate achieved great accomplishments in the end, while those who were born with a silver spoon in their mouths turned out to be not promising.

Pleasures and sorrows: pleasures bring forth sorrows; sorrows give birth to pleasures.

Joy and grief: in time of joy, one should prepare for grief because extreme joy begets sorrow; in time of grief, one should think of joy since joy makes people look forward. When you are having your day, something bad may happen, near or far.

Fatigue and idlesse: fatigue and idlesse interact as both cause and effect. They are the basis for each other.

Everything and nothing: nothing comes form everything, while everthing is the source of nothing. They are the basis for each other. "Being" equals to "not being", and "not being" is "being".

Good and bad: the two are one thing. One can only tell something is good or bad when he take human factors into consideration. Good things possess bad factors which may bring forth something bad, and bad things possess

① Quoted from *Lao Zi*.

good factors which may bring forth something good. Though someone is said to be kind, one still have to remember the possibility of evilness in him. And if someone is said to be evil, we also have to consider the chance of kindness in him. It is hard to guess a man's mind and it is difficult to judge whether he is good or not.

More and less: The more is the less, and the less is the more. An old saying on dinning goes: "One more bite may cause your lose weight."

Favors and frustrations: they give birth to each other reciprocally. What is disadvantageous to us is also favorable to us. All of us have a nature of laziness. If we fail to push ourselves, we would not be improved, let alone a rapid progress. Some moderate difficulties are good to our growth and we should think of difficulties in time of easiness. Difficulties are treasures to us. As long as we don't break down and still have hope, victory will wave its hand to us. Difficulties make people mature; pains make people strong.

Friends and enemies: friends may fall out and become enemies due to some factors. However, because of mutual interests, enemies may become friends as well. "People treat you the way you treat them."

Being fast and being slow: the two can also change into each other reciprocally. "More haste and less speed."① Moderation between the two is the best.

Honor and disgrace: honor and disgrace can transform to each other. Han Xin② once crawled between others' legs, thus gaining the impetus to growing stronger. After he succeeded, he even returned to his native place to thank those people who had humiliated him. Humiliations enable man to make reflections and be aware of himself. Man should remain indifferent to favors and humiliations. "Not pleased by external gains; nor saddened by personal losses." People should remember honor and disgrace at the same time.

Being good and being bad: the two can change into one another reciprocally. Beautiful persons, especially beautiful women, always being satisfied with external beauty and depending on it to live, are mostly characterized by inability. On the contrary, people who are not pretty work and learn hard to become capable persons, proving themselves in personal careers and specialties. And the disabled have a stronger will.

---

① Quoted from *The Analects of Confucius*.

② Han Xin(231—196), a famous Chinese military strategist in the Western Han Dynasty.

When in bad moods, we should think whether it is necessary to be looked happy. When being unwilling, we should think about the necessity of being assumed willing. When getting mad, we should think whether or not we are considered cool by others. We'd better think twice before we want to speak something. Those pairs—good things and bad things, giving and taking, safety and danger, are all interchangeable internally. As an old saying goes: "Living in the consciouss of suffering, dying in ease." ①

6.2.2 Make the most of the antithesis method

All things coexist with their respective antitheses. If one thing's antithesis disappears, the other thing will be no longer in existence. When taking antitheses into consideration, things can work best.

We should think differently by taking antitheses into account. Thus we can be relieved instead of worried, and delighted instead of sad. There are many pairs of antitheses. When meeting the former, we should think of the latter at the same time, for instance, advance and retreat, credit and punishment, honor and humiliation, health and illness, richness and poverty, happiness and misfortune, joy and sadness, benefits and losses, safety and danger, promotion and demotion, teasing others and being teased. Vice versa, when meeting the latter, we must think of the former. We should remain calm when dealing with urgencies; take your time when coping with something. We should take serious matters easily, and take easy matters seriously.

When you are praising others, you may also think of their bad sides. Fortunes do not come freely. So, if some lucky things come to you suddenly, you should watch out for them as they may be pitfall. Also, when you feel something bad happen to you abruptly, you can hope to be out of danger. Being nice to others might not bring a good luck to you; however, if you treat others in a bad way, there must be a bad luck to you. When you say good words for others, it may not always bring him good luck; when you say bad words for others, it may help him to achieve aim sometimes.

Confucius once said: "there are three things which the superior man should guard against. When the physical powers are not yet settled during the adolescence period, he should guard against lust. When he is strong and full

---

① Quoted from a Chinese classical literature *Meng Zi*.

of vigor during the postadolescence period, he should guard against being too aggressive. When he is at the topographic old age, and the animal powers are decayed, he should guard against covetousness."[①] When dealing with the strong desire of "lust, aggression and covetousness", man should regard their antitheses and weaken those desires, which will do good to his health. Confucius also said: "a man of great wisdom should appear slow-witted; a man of great successes should appear modest; a man of great courage and strength should appear timid; a man of wealth should appear unassuming. This is humility over humility." Those words of Confucius all aim at the use of antitheses.

Taking antitheses into account while coping with things will bring people unexpected, constant and extraordinary victories. An adept of antitheses will benefit you much. Man should keep "antithesis" in mind and put it into practice to explain things and instruct actions. All things go well or bad, and things which end up neither better nor worse will have no effect on us. Therefore, as long as we keep antitheses in mind, our thoughts will be comprehensive.

6.3 Risks come from puniness; hardships come from complexity

All risks in this world come from puniness. Risks emerge when children confront adults and a small country confronts a big one. If you have little fame, you will be held back by constraints everywhere. If you have little financial resources, you will have the risk of being unable to make a living. When you are on a low position, your words are of little effect and everyone can squeeze you, which is risky. A man who is as high as a hill would never have the risk of falling out from steep slopes. The survival of enterprises is not easy due to the risks coming from business operations. A small enterprise is like a boat in the sea and is of great possibilities of shipwreck. However, if it is a super ship, the factors of safety will increase a lot. And if bridges that extend in all directions are built across the sea, the risks brought by the sea will disappear. If joint management can be realized within a certain area or a country, the risks of business management will decrease. Supposing that the whole

① Quoted from *The Analects of Confucius*.

globe can work together, observing the overall situations, making reasonable plans, and producing and managing products according to plans, all business over the globe will be free of risks. In the event of the global unification of currencies and credit rules, there will be no regional financial crisis. Nowadays, different people set up a separatist regime by force of arms; many people are threatened by growing crisis. If the world can be united, danger will be eliminated. There are three ways to deal with dangers: avoiding danger, defusing risks, and growing stronger.

The bulkiness of barriers and the complexity of situations give rise to all difficulties in this world. The difficulty of solving the poverty problem of the world lies in the huge number of poor people. The difficulty of solving the food shortage problem lies in the considerable amount of food needed. Moonfall is hard because the distance is extremely long. Biology gene spectrum is difficult to describe due to its complexity. Discomposing difficulties and clearing our mind can assist us to solve difficulties.

6.4 Conservation of energy

The amount of things is limited and the gross is certain and unchanged. So man should keep a proper consumption of them because overspending will bring about high interest payments. For examples:

Drinking: an alcoholic drinks quite a lot of alcohols in his life. But the total volume of his lifelong drinking is certain. If he fails to control his drinking amount at the beginning then too much alcohol will spoil his health and he will be unable to drink thereafter.

Eating: if one eats properly, he remains appetitive; overeating makes people bored and force feed make people sick. People will not like to eat the food they ate in times of illness or the food that they had eaten too much, because the total volume is used up.

Energy: a moderate use of energies will make people in high spirits, and an excessive use of the brain causes a double consumption and does harm to health.

Love: a proper love can last permanently. Indulgence or solicitous service is counterproductive. Inordinate love generates hatred.

Lot: in the case that you and someone feel you two can hit it off and keep a very intimate relationship, you two will break up after a long period. But if a proper distance is maintained between you two, your lot will remain for a long time. "A hedge between keeps friendship green."①

Normal life: there are people who are arrogant and think high of themselves. But those people will have to answer for their conceits one day because they used up the corresponding volumes of their energy at an early stage.

Therefore, no matter what you are doing, you should ease off and never overdo things.

Everything in this world has its limit; an increase of consumption causes a decrease of the surplus. A German scientist Peter Axt and his daughter pointed out the theory of "life energy" after painstaking researches. They believe that everyone's life energy is finite and the speed of life energy consumption determines one's life span. For instance, constant work shortens one's valid work years because too much workload and tension will consume more physical forces. Too much exercising, especially when people force their bodies to exercise reluctantly, makes people short-lived. If we get tense, our body will have to endure heavy work and consume extra energy, which surely makes people ill after a long period. Hence, whether we're working or relaxing, we'd better ease ourselves to extend our life span.

Approximately there are three ways of energy consumption.

(1) Normal consumption: it is a moderate way, it leaves some leeway to last our energy.

(2) Luxuriant consumption: it is a superfluous way to spend double efforts than the moderate consumption.

(3) Extreme consumption: it equals to offset, running out of your energy in a short time.

6.5 Layers of society

Based on man's life quality, human society is divided into five layers.

(1) Threatening existence. One's personal existence is not guaranteed during the period of primitive society, during a full-scale war in a country or during a natural disaster, etc.

(2) Safe survival. Generally, people's personal safety can be guaranteed

① Quoted from a Chinese classical literature *Zhuang Zi*.

and people have enough to eat and wear.

(3) Affluent life. People are well-fed and well-clothed, having favorable living environment and conditions.

(4) Life of ease. Living in a productive society, people just enjoy a life of ease and happiness.

(5) Value realization. Almost everyone in society can realize their own value. The whole society has maximized all its positive values, emancipated the productive forces in the most effective manner, made full use of people's wisdom, utilized its resources most economically, greatly improved people's living environment, and enabled man to lead a life of excessive pleasure.

The criterion to judge the societal level is the quality of the 80% societal members' life.

A country, even the world, is in a constantly changeable social layer. A higher level today might become a lower lever tomorrow.

A specific time and an objective reality give birth to a certain social formation and layer, and the later then generates a definite human behavior which promotes human society in a specific direction.

6.6 Miraculous paradigm

Paradigm is a cognitive activity that analyzes, synthesizes, judges, and ratiocinates things on the basis of appearance and concepts.

6.6.1 Features of paradigm

Paradigm has many features: succession, inertia, tendency, multi-direction, and so on.

(1) Succession feature of paradigm. Everything in the world last through a form of succession: material transportation, work flow, human history, traditional culture, world civilization, continuation of the species, etc. It is self-evident that everything undergoes the processes of emergence, development and extinction. Every process is a constant phase of a thing's eternal development. However, every phase has its inevitable limitations. And they can only be broken in a form of succession to realize things' development.

Human thoughts also have the feature of succession: discussions and debates can develop and upgrade intelligence. Two people can have the same opinion for the same issue. However, if one of them presents it verbally or in writing, the other may immediately put forward a further perspective based on it. This doesn't mean that the later is smatter; on the contrary, it is the succession feature of thoughts. As a proverb goes, "Debates make people closer to the truth." Why does it make sense? On one hand, it is because man concentrates most attention in times of discussion and thinks very quickly; on the other hand, it is partly due to the succession of thoughts. When you put up a question, all your mind will be occupied by it, which stops you from thinking further. And at this moment, if you hand your already processed question over to another person who has no burdens of prior thoughts and can conduct full brain work, he will take the question to a higher level. Thus, during the process of succession, a problem will be readily solved due to doubled solution efficiency. Similarly, if we write down our opinions, we will have a deeper understanding after some time or exactly after we finish writing.

(2) Inertia. A person's way of thinking and thinking results are often the same with what he did last time. Only when he has learned some new knowledge or profoundly corrected his former thoughts can he change his previous ways of thinking and thinking result. Hence, people have to constantly study and correct the wrongs to make their thoughts incline to right. For instance, students have to study harder and correct themselves again and again to put their ways of thinking right; people dream about what they think in the daytime; if a tendency haunts us last time, it will make us yearn day and night this time.

The inertia of thoughts can be divided into two types: one is self inertia, namely habitual thoughts; the other one is guidance thought which means a man's thoughts are provoked by other people or objects, such as the publicity of murders, horror movies, deceitful tricks, others' exemplary practice, etc. So, if we don't want other people to have a certain thought, we'd better never bring them up.

The actions of a person at every turn and the impacts the exterior have on him will change his cognition, personality as well as likes and dislikes. Three times of experience will turn into a habit.

(3) Tendency. People's thoughts have tendencies. Once they meet an idea which impresses them, makes him worried, anxious or puzzled, or forming faultage thinking, they would keep thinking about it unremittingly. People's confusion would continue to increase, which may finally make people unable to extricate himself and get some psychological illness.

One way to prevent a man from thinking about that is to straighten out his thinking or replace that idea with something more stimulating and interesting. The stimulation can be physical or a stimulating suspense, which includes making an implement, taking any associative subsidiary measures, and so on.

(4) Multi-direction. Thoughts can develop from a multitude of perspectives. Thinking about a thing can start from perspectives of time, space, cause, result, order, start and finish, the beginning and the end, complexity, size, speed, motion state, advance and retreat, associations, internal laws, external factors, self influences, outside influences, analysis, synthesis, concentrative thinking, divergent thinking, and so on. As one's thoughts develop with depth and breadth, they will be generalized to philosophy category with the specific becoming the general. People should develop the capacity and the habit of thinking from all sides. Intelligence lies in one's ability to think wisely and deeply from a small clue.

It is said that, in the early years of the period of National Government, Gu Weijun (Vi Kyuin Koo) took up the post of the ambassador to America. Once, he joined an international dancing party in Washington. and an American lady was dancing with him. Suddenly, she posed a question to Mr. Gu: "Which do you like better, American lady or Chinese lady?" Then Mr. Gu replied with a smile, "I like the lady who likes me. " This brilliant answer is of courtesy and respect, avoiding the awkward situation of attending to one thing and losing another.

When we are meeting the situation that the subject can not find the object properly, we can make a detour by letting the object choose the subject. Here are some other examples. If someone asks you "which song do you think is better" and you like both, you can replace the subject "I" with "you" and reply "I like the one you like". Otherwise, you can seek some other ways by changing "like both" into "like parts", answering that they both have distinctive features. Therefore, one should do more exercise on changing elements of sentences to gain wisdom and improve response capability.

(5) Limitations. It is hard to do things or develop an opinion, while it is easy to find faults. This, on one hand, reflects the successive feature of thoughts. The fault-finder usually has less spiritual burdens and pressure than people who do things, and he can concentrate his attention to pick holes. On the other hand, it states that any truth has room for improvement and everything has the possibility to be improved. The objective is more complete than

our knowledge and what happened is more than what we knew.

(6) The otherness feature of thoughts. Differences exist in human thoughts. First, the thinking subject varies. That's why people have various thinking methods and results. Second, thinking subjects are in different objective situations. As the relative objective condition changes, the thinking method and result become different. A single phenomenon can always have many interpretations. "The benevolent see benevolence and the wise see wisdom."① Therefore, any two people couldn't have totally the same thinking method and result, which explains people's differences in actions.

6.6.2 Thinking level

According to man's differences on thinking level, man's thinking can be divided into plane thinking, stereoscopic thinking and transcendental thinking.

(1) Plane thinking. Normally, most people have plane thinking, or secular thinking, which is inflexible, traditional. This kind of people may treat others with deceits or goodness. They may also seek quick success and instant benefits, or give tit for tat. They only know to go from one to two, having no idea of the possibilities of "one multiplies two" or "one is divided by two". The deeper connotation can't be perceived or thought of by them, which is often a result of habits or heredity.

(2) Stereoscopic thinking. It is superior to common feelings and emotion. People thinking stereoscopically do things with reasons. Their thoughts are based on the overall situation and have impacts on the whole world. It is the way the leader thinks.

(3) Transcendental thinking. Saints' thinking belongs to this type, exceeding ordinary persons', taking on all the features of the world. For instance, Jesus instrusts people if someone strikes you on the right cheek, turn to him the other side also. When a roman soldier thrust a lance into Jesus' side and crucified him, he even begged for forgiveness from God for the solider because he thought those soldiers didn't know what they were doing.

6.6.3 Foundations of thinking

Man's thinking is an automatic process based on consciousness. One's

---

① Quoted from a classical Chinese book *The Book of Changes* (*I Ching*).

foundations of thinking determine the thinking result and behavior. Foundations of thinking are not accomplished in one action but need days and months' accumulation. If one is polite and well-mannered at usual, he will be the same when he encounters something or thinks about things. After a person assumes office, constantly commanding others and being flattered, he would become arrogant naturally and put on official airs. If we look down upon someone in peacetime, our thoughts and actions would do the same when we are making contact with him.

After neatening, applying, reorganizing and creating the basic information in the mind, human being get thoughts. There is no thinking without basic information. Ample basic information gives birth to great original thinking. Inspirations are the fruits of thinking, caring only the people who are familiar and skilled with their career. Human brain is the most accurate and mysterious computer. It's better to consider that everything has a worth and not to limit yourself to a specific development direction and area set by yourself. Our brain is a superlative computer that receives and operates information. All the information we have received is useful and nourishing to our brain. The more diversified the received information is, the more accurate our thinking can be. Also, it makes our career more of a success.

Completing some tough tasks does not require much brainpower. People are born with great differences in intelligence. However, most highly intelligent people often fail to make an achievement, even in researches and inventions. The reason is that man's brain has great potentials. Although someone is not as smart as others, their brainpower is enough for use. Intelligence is one side of the coin of success, while the key is whether or not a man deals with a concrete matter with perseverance. A leader or a president, however, is not the smartest and the most flexible person, but a man of justice and thoughts.

Thinking has foundations. Discoveries, inventions, and inspirations are results of human thoughts. The subject or thinking method one is expert in during one's childhood does not change as time passes by. The original thinking mode still exists. If one can have them carefully arranged, one will succeed. Human thoughts are formed based on one's background, experiences and orientation of choices.

6.6.4 Measures to develop thinking

There are many ways to develop one's thinking. The followings are some examples. Firstly, sleep is the filling station for thinking. Human brain is an extremely complex and miraculous organ and sleep does good to it. Modern science has proved that our brain still works in times of sleep. In different phases of sleep, our brain classifies, files, and calculates different information. The calculation result reveals itself in dream or in the morning of the next day. Most people's sleeping hours make up about one third of their whole lifetime and one night's sleep can be divided into five phases, with each phase playing different roles in helping us organize our experiences, consolidate memories and collect thoughts. Lack of sleep severely affects man's thinking ability, learning effects and work efficiency. Secondly, physical exercise is a roll booster for thinking activities. They make people lucid by improving blood dredging and providing people with nutrients. Thirdly, inner peace is the best surrounding for thinking. "Still waters run deep." Because some people are not interrupted by the outer world, otherness, or innermost beings, they are able to own inner peace. Hence, if one wants to be calm, one has to separate "otherness" from objective environment and one's innermost being. Fourthly, activate thinking. People have a nature of laziness. People in depression won't have a sharp mind. As an old saying goes, people show resourcefulness in an emergency.

6.7 Generalities of things

Things are unity of both inside and outside like a steam bread. The appearance of steam bread may be the same but its inside may be different, and vice versa. What people see is only appearances of things, but there are reasons for those appearances inside and deep side.

The existence and developing laws of things are decided by the nature of objects, while their nature is determined by its internal structure. "Adaptation" is the summarization of all laws whose essence lies in "adapting to".

The development of all things in the universe is driven by effects and it is inevitable that effects determine everything in the world. Social relations generate all things in society. All the social relations are based on production relations, i. e., economy relations and all social relations change according to the economic relations. All problems of society certer on man and how to deal with human beings, while the whole efforts of man concentrate on how to live a life.

Everything is composed of two opposite sides. The sole direction things develop towards is their opposites. The only matter that differs is the duration a thing stays on its one side. The development trajectory of things is like the tracks of feet, moving right and left alternately. There are two kinds of things in the universe: one is the balanced, the other the imbalanced. The balanced will turn into the imbalanced sooner or later, and the later will do the same. A thing's two sides will switch into each other. Only when people realize that point can they live a life of contentment and happiness.

The final structure of any substance is the same no matter how constantly the substance is decomposed. And if the final structure is decomposed, the substance turns to non-substance. The universe is an unlimited field where substance and non-substance interchange with each other. The ultra result for spirits which keep developing positively is also the same, which is "adaptation". All forms of substance in the universe mix with one another. Human world is the mingling part of "adaptation" and "non-adaptation".

6.8 Exploration of the universe

The universe is mysterious, which causes innumerable guesses. The followings are the author's personal opinions.

6.8.1 Substance and non-substance

At the beginning, let's discuss some concepts.

The extinction of celestial bodies. Since a celestial body came into being, it goes through the change from prosperity to recession. The outer shell of a celestial body is supported by its inner nuclear reaction. When the energy produced by thermonuclear reaction can no longer prop up the huge outer shell because the central fuel is almost drained out by nuclear reaction, the core of a celestial body starts to collapse and shrink under the great pressure of its outer shell; at last it becomes extinct. Any celestial body is doomed to die out. To the entire universe, a celestial body is nothing but a grain of sand.

The formation of black holes. A black hole is easily to be taken as a big black hole literally. Actually it's not. The so-called black hole is a kind of celestial bodies: its gravitational field is so strong that even light cannot escape from it.

Gravitation field can make space-time crooked according to the general relativity theory. A fixed star's gravitation field barely has impacts on space-time if its volume is huge. The light from any spot on the surface of a fixed star can irradiate in any direction straightly. The shorter a fixed star' radius is the greater influence it has on the crooked space-time around. The light beamed from some angles will go back to the surface of the fixed star along the crooked space. When the radius of a fixed star becomes small and reaches a certain value (astronomy calls it "Schwarzschild radius"), even light beamed perpendicularly will be captured. At this time, the fixed star turns into a black hole. We describe it with "black" because it is like an endless hole in the universe. It seems that everything that falls in it can never escape.

When a fixed star is aging, the whole of it will irresistibly shrink toward its center until its volume tends to be zero and its density tends to be infinite. And when its radius shrinks to a certain value (Schwarzschild radius), as mentioned above, giant gravity makes nothing able to come outside, even light. Then the connection between the fixed star and universe is cut off—a black hole is born.

Substance and non-substance. Substance is what can be touched and be seen by human beings or what could be observed by equipments. Non-substance is what man can't capture. Substance and non-substance are reciprocal and can transform to each other. Non-substance is the origin of substance of the universe. Substance will turn into non-substance finally. Small celestial bodies and mass space mess substance (relative to celestial bodies) transform to non-substance after being totally smashed by the "super smashing machine". Human beings and the tools they use are from substance, so that non-substance could not be observed. If substance is infinitely disintegrated, all parts of it would become non-substance.

6.8.2 Universal circulation

The universe is circulating. It is in a constant circulation in which substance and non-substance transform and alternate into each other. Non-substance forms substance through aggregation. When substance increases and squeezes to a certain degree, it would explode and form all kinds of celestial bodies which make up a galaxy centering fixed stars due to their gravity. Celestial bodies are constantly aging and dying. They are continually disintegrated into non-substance. Non-substance aggregates and turns to substance again. While the area of substance keeps expanding, non-substance uninterruptedly keeps pace. That's how the universe circulates. But the universe is endless. Therefore, the two processes of circulation do not happen at the same time (the universe can't be purely substantial or wholly non-substantial). They go on alternatively. The universe has no origination; only some part of it is relatively born and only content and structure change. The universe would not die; only some part of it would relatively disappear. Substance area, non-substance area, and empty area which is totally empty make up the universe.

It is reported by the Xinhua news agency in Washington on August 25th that American astronomers recently discovered a super cavity with a diameter of about billion light years in the universe. There are no celestial bodies, no galaxies and even no mysterious invisible dark substance in it. Astronomers feel strange about its emptiness.

Lawrence Rudnick, an astronomy professor in the University of Minnesota stated: no similar cavity was discovered before. And we have never expected to discover such a big cavity.

Astronomers have already known the existence of cavity in the universe, but the scale of the latest discovered one is still out of their imaginations.

Rudnick and his colleagues said in the latest *The Astrophysical Journal* that They discovered the cavity in Eridanus which is on the southeast of Orion while they researched on the observed data from the radio astronomy observatory of America and Wilkinson Microwave Anisotropy Probe satellite.

Rudnick said that this cavity is different with others in area. The temperature of its cosmic microwave background radiation is a bit lower than that of other parts of the cosmos. At present some researchers call it " cold zone". The difference in temperature perhaps can only be explained by stating that "There is an 'empty hole'." Astronomers will continue to verify, this view.

Some other astronomers see with Rudnick's standpoint, Brent Turi, an astronomer from the University of Hawaii in America said, "it seems that its contents are taken away." He deems the reason why "empty hole" exists is that substance of this zone has been attracted away by other zone which has larger gravitation.

6.8.3 The cause and effect of life

Life is sustained by things it derives form. All lives on the earth come from water and carbon; hence water and carbon are necessary to sustain life from the beginning to the end. Supposing that lives from other planets don't come from water and carbon, their lives will depend on other things which make them instead of water and carbon. Consequently, the origin of life can be deduced from the things which are needed to sustain life.

6.9 Real God

6.9.1 The types of gods

There are three types of gods in the universe, divinities, sorcerers, and spirit. Gods are imagined by people, sorcerers are pretended. Only spirit really exists in the world.

Divinities fall into three kinds: the first kind is the creator and governor of all living things in the world and exists in the mind of the superstitious; the second kind is the soul of dead people who possess extraordinary abilities and high morality; the last kind is people who have super powers, and immortality. In the early developing stage of religious belief when the system of clan existed, every clan had its own god to worship. As the system of clan developed, people began to worship several divinities, and every divinity had his own name, specialties and activity area. After countries came into being, one country chooses one divinity to embrace. And when emperors emerged, the concept of king of divinities appeared. As Engels wrote, "the king of divinities" is merely a duplication of a monarch. No king, no king of divinities." In modern religion, people worship several different divinities.

People believe in divinities, but they can not see or touch them, thussorcerers rise in response to the conditions. By pretending to be incarnate divinities, sorcerers deceive people by making them believe they can see and touch divinities. A sorcerer's job is to pray to divinities for people by affection. However, their deceitful tricks can be totally disclosed with the help of science. For example:

(1) Sacred fire. Sorcerers sprinkle a mouthful wine towards a bamboo hat and make it self-ignite. Then they make up lies and deceive people. Actu-

ally those sorcerers have added some white phosphorus to the wine. White phosphorus has a low ignition point. After the wine is sprinkled and rubs air, the white phosphorus self-ignites immediately and ignites the wine.

(2) Killing "ghosts". Sorcerers first make a paper man with the paper which has been dyed by turmeric and cut the paper man with a knife that has been dipped into the water containing soda. Then part of the paper man immediately gets red, which is commented by sorcerers as "the ghost has been killed". But the red is ascribed to the chemical reactions of alkaline water and turmeric.

(3) "Sealed books". Sorcerers write characters with the writing brushes having been dipped in the cooking liquor containing starch (like rice-water) ahead of time. Those characters become invisible after the water evaporates. And if people spout sulphur solution on the paper, those characters will become blue. The phenomenon comes from a simple chemical reaction. The cooking liquor can be replaced by alum water, milk, etc.

(4) Capturing ghosts in an oil-bath. Sorcerers add a little oil in the oil-bath as well as a lot of vinegar or glauber's salt. Because vinegar has a low boiling point, it comes out in a gaseous state when heated up. It seems the oil is discharging smoke. Actually, the oil is not hot at all. Sorcerers' hands won't be burnt in the oil when they are capturing "ghosts".

Religions, doctrines, thoughts and so on are all spiritual products created by heroic individuals and communities to lead people to develop towards a certain direction. They are improved by generations and generations. Religions play a part in unifying people's thoughts, adjusting people's mood and persuading people to be kind, but they will lead people astray at the end because of their unscientific nature. Spirit is a real god, one's own god, and the most and only reliable god. Body and spirit are closely correlated, sharing joys and sorrows.

6.9.2 The relationship between body and spirit

Human beings are the combination of body and spirit. Body and spirit are two systems, having their own developing conditions and regulations. The basic for the appearance and development of body is nutritious substance, and for spirit it is spiritual substance. Spiritual substance is things that produce spirit, including objective reality, oneself, memory and so on.

Spirit is the activities and outcomes of people's consciousness and thoughts. The essence of spirit is substantial, but it has different forms: spiritual activities are not easily felt. For example, memories are spiritual, but they are all prints printed in some parts of our brain. In the future, people can gain knowledge without learning if some special substance is put into the brain, which proves that spirit is also a kind of substance, a special form in substance's existence. Spirit has two kinds of the existence: spiritual activities and spiritual outcomes. Spiritual outcomes fall into two groups: one is exporting (including the bio-waves and works released when people think), the other is internal. The internal disappear as spiritual activities end, but the exporting never dies.

Spirit and body rely on each other. Body is the house, carrier, weapon of spirit and the executer of orders. Spirit isthe support, soul and governor of body. Body dies as spirit dies, and vice versa.

Spirit and body influence each other. Body depends on spirit to be vigorous; spirit functions through body. Spirit becomes high if body is healthy; body will be strong when spirit is high. Downhearted spirit makes body exhausted; tired body makes people out of spirit. A man can't be considered healthy unless he is sound in body and spirit at the same time. A man has to exercise his body, but his spirit needs more exercises. Spirit cultivation is more important than body cultivation.

Spirit is superior to body. Spirit is the governor and body is the servant. Body can be disabled while spirit is sound. However, body is doomed to break down if one's spirit is chaotic. One's body becomes a walking corpse if his spirit is controlled. Because of belief, Islam extremists should have the courage to make suicide attacks; because of the belief in communism, soldiers could devote their lives fearlessly.

6.9.3 The huge role of spirit

Men have great potentials which can only exert influence when inspired. Inspiration refers to the inspiration of spirit. Spirit releases its limitless energy through body. It controls and orders body. Energies from body are limited but the energies from spirit are unlimited. Spirit functions through body.

(1) The positive and negative effects of spirit. The influence of spirit can be positive or negative, which depends on how you use it.

The power of spirit is huge. Spirit produces great energy by instructing body. The energy from spirit is so great that it is usually inconceivable. Body is the weapon of spirit and spirit is the governor of body. One who desires to make achievements must possess the spirit of self-improvement.

Objective things can only be correctly reflected by proper spirit. Goals have to be realized by the body led by spirit. Spirit functions for controlling: sometimes one thinks oneself great, and sometimes futile. In fact, body stays unchanged; spirit denies or affirms oneself.

Long-term spiritual states such as nervousness, anxiety, hostility and depressions can cause diseases in nerves, heart, blood pressure, and so on. Michael Flora, the cardiovascular medical professor in the University of Birmingham, said that depressions can at least double the risk of heart attack of a healthy man. Hostility is also one of the causes of heart diseases. Duke University in America once did a large-scale survey on patients. According to their survey results, strong hostility increases patients' risk of dying of a heart attack by 29 percent; however, the percentage exceeds 50 percent among people at age 60 or under. On the contrary, if one's life is in a peaceful state, his blood pressure and blood sugar will become normal gradually, reducing the chance of getting ill.

Divinities, works and incarnations of spirit, are imaged. How much a man believes in divinities (or God) can tell his mentality, and the stage of life (including "vigor" "steady improvement" "resigning oneself to one's fate" "being downhearted") he is in. Children and teenagers don't believe in divinities. Even if they do, it is just out of curiosity and mystery. They are not superstitious.

Spirit and body have their own extremely complex physiological mechanism. Both of them can regulate, adjust themselves, and function properly; they can also disturb and destroy themselves. Many diseases can be cured from the aspect of emotion. Psychological treatment is cheap and void of pain and risk, which should be advocated.

Think first when you meet a trouble. The effect of thinking is greater than that of body. So it is wise to think first and silly to act first; thinking first saves time, and acting without thinking wastes time; spirit functions faster than body. A precise stratagem brings a big victory; a small scheme brings a tiny victory; no victory is wined without plans. Spirit is always stronger than body. We should use our head more and run errand less.

Think before you leap. Spiritual behavior such as motivation, will, and plans are like gun sight. A little error in ordering body may lead to a large discrepancy. So you can never be too careful before you act.

(2) Spiritual punishment is crueler than physical punishment. Human beings are spiritual animals so that they can be punished physically and spiritually. Physical feelings can be easily forgotten. As an old saying goes, nothing dries sooner than tears.① Spiritual feelings are unforgettable forever. One becomes worried when thinking of something terrible and becomes happy when thinking of something pleasant. Physical feelings are usually connected with spiritual feelings but more often than not physical feelings are stored in spirit.

It is reported that eight years ago Turner hit a young girl named Susan who was still in high middle school to death with his car after drinking. At that time, Turner accepted the punishment come up with by Susan's parents: every week Turner had to send a check to Susan's parents, but the check must be written to Susan. Every check values one dollar—just the right amount—only one dollar. What is more, the check must be sent every Friday in the following 18 years.

Turner thought it was dirt cheap. One dollar every week of the following 18 years only amounts to 936 dollars in total.

Eight years later, Turner could not stand it and did not send checks on time any longer. Susan's parents sued him again. Turner was on the edge of breakdown. He told the judge with tears, "I really can not stand it anymore. Every time I wrote Susan's name, the sense of guilt overwhelmed me. Susan's death is still vivid in my mind. My wound is so deep and it has to be torn once a week. How can I live in the following long ten years? Maybe I will be mad in no more than ten years. I used to have a bee in my head when lying in bed. Now whenever I lie down, I see Susan walking towards me ..." He asked to compensate more and pay all the compensations at a time.

His demand was certainly refused by Susan's parents and the court. His spiritual punishment continued.

If Turner was punished to give a large amount of money at the outset, he would soon become easy and forget his sin because he begrudged the money he compensated. Only by experiencing such spiritual punishment could he truly realize that no matter what kind of punishment could change the evil consequence caused by him.

---

① It means "Once on shore one prays no more."

(3) Spiritual behavior needs to be corrected. Spirit can exert both positive and negative influences on body. Negative spiritual behavior need to be corrected timely. For example, chronic spiritual experience. Experience consists of physical and spiritual experience. Spiritual experience refers that a man is willing to think over the exciting things he met, ponder things that happened to others to assimilate their experiences, and fantasize things in the future. This kind of experience can stimulate both body and spirit: spiritually experiencing falling from a high place makes hands sweaty; experiencing being drowned in water in mind suffocates a man. If a man forms the habit of this experience by constantly experiencing disasters in mind, his body will secrete various kinds of harmful substance which do harm to both physical and mental health and make people get ill and senescent.

Similarly, perfectionism may be a big enemy to spiritual health. Since there is nothing perfect in the world, importuning things to be perfect will definitely cause spiritual burden which one can't cope with.

Human beings are depressive animals. One has to release his depressions when depressed. Humans are instinctive to save themselves. Releasing depressions is one of those instincts. Different people have different ways to release depressions. Artists tend to liberate themselves by daydreaming—creating works to release their feelings. This phenomenon is quite common if we make a general survey of the history of literature home and abroad: Qu Yuan's *Lisao*, Sima Qian's *Historical Records*, Pu Songling's *Strange Stories from a Chinese Studio*, Kierkegaard's *Deadly Macronosia*, Hugo's *Notre Dame de Paris* and so on. There is no doubt that these works are created when authors vent the anguish of their hearts and redeem themselves. Therefore, anguish is not totally unfortunate for artists, because great works more often than not come from great pain. "A poetic masterpiece through the ages definitely appears when the poet is frustrated. Only the words of a frustrated poet can last forever." (*Poetry Collections of the Qing Dynasty*) Singers release their feelings by singing songs and athletes by doing sports .... For most people, physical labor and exercises are necessary for health. Singing and laughing are care, renovation and improvement to spirit.

(4) Human beings live on spirit. From the physical aspect, people's five sense organs have their own spiritual and physical functions. Our eyes show the world to body and absorb nutrition for spirit; our mouths take food for body and meanwhile convey the contents of spirit by speaking; our ears identi-

fy sounds for body and receive messages for spirit; our noses breathe for body and sense all kinds of smells which can also stimulate spirit. All people like to look and listen to satisfy their spiritual needs. Spirit can not stand loneliness, and we must satisfy its needs to look and listen. Spirit supports body and can make body tolerate limitless pain and overwork. However, once spirit breaks down, body will do the same no matter how strong it is. If a man has some spiritual problems, he will suicide no matter how good his life is. But a strong spirit can survive a man who has been trapped by an earthquake for ten days.

From the aspect of emotion, people will feel good about others' words and behavior if they get on well with each other. Once their friendship is gone, they will hate each other even for the same words and behavior. A bad mood makes people weak and unable to coordinate actions. "People are in high spirits when involved in happy events." High spirits brings man health. They make people feel high and noble, otherwise the opposite. Even though the surroundings are not changed, a good mood makes them better and enables man to feel in the pink. Beauty springs out of heart, while happiness is of the most beauty.

From the aspect of life goals, everyone has his beautiful expectations or great goals in their youth. Some people's goals are degenerated by time, which makes them unconfident and desperate. However, some, in the same situation, get stronger in front of setbacks. The key is one's spirits. People's spirits determine goals and fill people with confidence to succeed. They also assist people to comprehensively analyze the information at hand and make thorough plans. And when people are in the course of achieving their goals, spirits help them find their willpower. So, all through the realization of goals, spirits design all and direct the body.

From the aspect of society, man should have a fighting spirit. Without it, man will become a walking corpse and running flesh in search of souls. Before founding the People's Republic of China, Communists sacrificed their precious lives to fight for communism; during China's national construction, heroes sacrificed all their conditions for a wealthy and strong China. Those Communists and heroes are all sustained by their fighting spirits. Pursuits of materials only belong to low-class animals, while the purists of spirits are noble and high-class.

Humanbeings are spirit animals, they rely on their spirits. They are apt to underestimate themselves because of setbacks and overate themselves in prosperity. "Nothing is more lamentable than a dead heart." And the motto of Tsinghua University "as heaven maintains vigor through movements, a gentle man should constantly strive for self-perfection; as earth's condition is receptive devotion, a gentle man should hold the outer world with broad mind"① runs over with the power of spirits. God blesses those who help themselves. A man's fate depends on himself, and no one can give you a hand. Here depending on oneself means depending on one's spirits. As the elves of the universe, people live on spirits. Controlling a man' spirit is the way to control him. And if someone wants to exterminate another man, that man's spirit has to be destroyed first. There are thousands of ways to control man's spirit, such as doctrines, thoughts, disciplines, laws, religions, and so on. As long as they can control spirits well, any way will do.

(5) Only spirits last forever. Those things that can touch people depend on some spirit. And those which can last generations are ascribed to spirits. Only the spiritual tings can live forever. The materials have to combine with spirit to be meaningful and be inherited. "Even if you have 10,000 hectares of fertile land, you can only have three meals a day; even if you have 1000 rooms in a mansion, you can sleep on half a bed one night."② Once a man is dead, nothing in the world will have something to do with him. Everything is gone in the wake of death, while only spirit is immortal. For instance, the reason why Alfred Bernhard Nobel, a Swedish chemist, engineer, industrialist, and the king of explosives, who established the Nobel Prize, can gain immortal fame is that he combined his career and wealth with his dedicated spirits. In his whole life, Nobel has been claimed for motive power from the nature under the threat of Death. He only used several short sentences to describe his lifetime scientific achievements: "The writer was born on October 21, 1833. He learned his knowledge from a family tutor and never went to college. He was particularly devoted to the study of applied chemistry. In his life, the explosives invented by him include: high explosive, smokeless powder, 'Balisiti' or Number C89. In 1884 he joined the Royal Swedish Academy of Sciences, the Royal Society of London and the Paris Society of Civil Engineers. In 1880, he was awarded the Scientific Medal issued by the King of Sweden and the French Badge." Before his death, he made a will in 1895, bequeat-

① Quoted from *The Book of Changes*.

② Quoted from a Chinese classical literature *Zeng Guang Xian Wen*.

hing most of his wealth (9,200,000 dollars) as funds and using its annual interests(200,000 dollars per year) to award academicians who made the greatest contributions in the five fields of physics, chemistry, biology, medicine, literature and peace(in 1969 Sveriges Riksbank added a prize in economics). And since the year of 1901, the prize will be awarded on December, 10th, the memorial day of Nobel's death.

6.9.4 The power of society spirit is the pillar of society

Society spirit refers to the common spirit owned by every member of the society. It is the fruit of social interests, caused by great spiritual inspirations. Society spirit has tremendous power to be invincible.

Every single man is a god. As the saying goes, Gods' abilities are remarkable. As long as a man can realize his abilities and dare to utilize them, he will have limitless creative power and make great achievements.

Spirit is the true and the only god. It is a proper backbone to human beings. Let's bring brilliant and powerful spirit into play to create a bright future.

6.9.5 Eulogizing spirit

Spirit is broad and profound. It stretches its scope fantastically. Spirit is the world. Spirit is the universe. All is in the spirit.

Spirit is a commander, able to let the mountains make way. All the seas and lakes abide by Spirit's orders. Spirit sweeps away all obstacles. Every living creature applauds to Spirit.

Spirit is a god, the only god between earth and sky. Spirit dominates the whole universe. All things and creatures surrender to Spirit.

Spirit is a designer,playing his cards right and gain a decisive victory miles away. Spirit makes the blueprint and charts the future. All should wait in hope.

Chapter 7 Adaptation rules the world

Adaptation is a natural law and living without adaptation will suffer retribution. If you are familiar with adaptation, you will be the ruler of everything.

Someone says knowing half the *Analects of Confucius* can rule the world. But I think the single phrase "adaptation" can rule the world. Things would be messed up are due to that they are not following the rule of adaptation. Everything will become tragedy without adaptation. There are two kinds of tragedies in the world: Selfishness is the root of all evil; not following the rule of adaptation is bringing misery to yourself.

First of all, I'd like to propose a question: Is there anyone who wants everyone to disagree with him? Definitely no one wants that. If there is someone who likes that, he must have some mental problem. Being smooth is a common desire for people, and the best thing in our lives is going smooth in everything. In the interpersonal relations, the common rule is: if you want other people to treat you the way you want, you have to treat them the way you want them to treat you. In other words, if you want things to go along your way, then you have to comply with other people. Thus, adaptation becomes a common principle. Well, is adaptation a universal principle for human beings? How do we accomplish adaptation? This is what we are going to discuss about.

7.1 Adaptation is a natural law

Adaptation is a natural law, and it is the principle of everything that exists and develops.

First of all, let's take a look at how the Chinese character "顺" (adapting to) is made up of. This Chinese character consists of "川" (rivers) and "页" (a piece of slice), meaning a piece of slice flowing along the river.

This piece of slice seems to have two kinds of fate to choose: floating along the flow of the water, or clinging to something it can depend upon, and finally being torn apart by the water.

An individual is like a piece of slice in the universe river, which does not have the power to fight against the universe, but to adapt to it.

(1) The natural world commonly follows the principle of adapting to, which is universal. The universe, according to gravity, is composed of several galaxies and objects, which "comply with" the gravity, for example, the earth goes around the sun, and the moon goes around the earth. Every object in the universe moves in accordance with the law of adaptation.

All natural phenomena on the earth are following the law of adaptation. Water always finds its own level; all rivers flow into the sea and the sea water rises and falls as a result of the gravity of the moon; water turns to vapor because of heat and then vapor turns to rain because of cold; air turns to wind because of different temperature and pressure. Earth adapts itself to sunshine temperature and then it can turn to four seasons; the stone can not climb up the mountain. The law of adaptation is the eternal law of nature.

(2) Human being generally follows the adaptation principle. It's obvious that nobody is going to disagree that the nature complies with the adaptation principle. Well, do human beings have to follow the principle? The answer is for sure. Everyone is a unique individual, and each has his or her own thought, which are different; and this is also a principle which we should obey. Those who learn to adapt will go lucky, and those who don't will fail. Adaptation is an eternal truth. If you don't learn to adapt, you will get bad luck.

Celestial gravity creates gravity force, making people able to live on the earth; if we do not comply with the principle and want to live without gravity force, we're meant to die.

The earth turns around the sun, day and night, season after season. If we do not follow the rules, can we really make a change? We cannot.

As time goes by, people grow older and older and it is impossible if we want to be young forever.

These are the basic laws that humans make a living by clothing, food, shelter and transportation and become energetic through sleep; here comes the question, is there anyone who would fight against the laws? No, people who are not crazy won't break the law, but try and manage to sleep well and eat well to follow the principle.

These examples given above are very simple and common, and everyone knows that they should adapt to the law, which means the adaptation principle is universal and right, and we should make good use of the adaptation theory to guide our action. Just as what Tolstoy has said: "The most profound truth is the most ordinary one."

(3) The historical verification of the rule of adaptation. After Yellow Emperor① (a legendary ruler), when Yao②, the ancient Chinese tribe's chieftain, got old, he summoned the tribe alliance meeting, discussing the problem of voting the successive chieftain. Everyone recommended the virtuous Shun③. After a long investigation on Shun, Yao abdicated and hand over the chieftain position to Shun. This way of alternating the chieftain was called "abdication system" in history. During Shun's time, there were flooding everywhere in Yellow River④ valley and Shun sent Yu⑤ there to prevent flooding by water control. For thirteen arduous years Yu traveled everywhere, using methods of channeling and dredging (adaptation methods) to prevent floods and finally succeeding in subduing the Great Flood. Yu contributed a lot in subduing the Great Flood and won the trust of the people. Later, Shun abdicated and hand over the chieftain position to Yu. The abdication system adapted to the social realities and people's wills at that time, So that period became the most prosperous times in the primitive society. Owing to the surplus of living materials, this period became the transition period from primitive society to slave society.

Adapting to the development of the productive forces and the private ownership at that time, China entered the slave society. The Xia Dynasty was the earliest slave society in China. In the 16th century B. C., the last emperor

① The Yellow Emperor, or Huangdi, is a legendary Chinese sovereign and cultural hero who is considered in Chinese mythology to be the ancestor of all Han nations.

② Yao is a Chinese tribal alliance leader in ancient times, creating the abdication system.

③ Shun is the late leader of the tribal alliance of the patriarchal clan society. Yao passed throne to Shun.

④ The Yellow River is located in the north of China. It is the most important source of Chinese civilization, and it is called the Mother River by the Chinese people.

⑤ Yu is the first emperor of the Xia Dynasty. He controlled the flood successfully.

of Xia Dynasty Xie Jie[①] was on the throne. He and the noble slave owners were leading extravagant and dissipated lives and performing tyrannical and brutal ruling policy, which were not adapting to the people's will. Therefore, Xia Jie was defeated by the chieftain Tang of the Shang Tribe who was living in the downstream area of Yellow River. The Xia Dynasty dies out and was replaced by the Shang Dynasty.

The slave society performed the "nine squares"[②] system (of land ownership in China's slave society) with one large square divided into 9 small ones (like the Chinese character "井"), the 8 outer ones being allocated to serfs who had to cultivate the central one for the serf owner. In the slave society, the slaves can do nothing but to work and all the products belonged to the slave owner. The slave owner would send a lot of workforce to supervise the slaves, which was inhuman and not adapting to the human wills. Therefore, by the end of the slave society, the slaves had gained the previledge to keep a little part of the products while handing over the most part to the slave owners, which was the exploitation method of feudal society. This method was more adapting to the reality of the society and more adapting to the people's will at that time and it was progressive at that time. In feudal society, the farmers' status was improved and they could enjoy more freedom, which improved the farmers' enthusiasm for production and promoted the development of the social production.

During the Northern and Southern Dynasties (420-589), Emperor Xiaowen[③] in the Northern Wei Dynasty (368-534) performed the system of land equalization, which means to allocate the land controlled by the government to the farmers equally according to the population. It greatly promoted the recovery and development of economy at that time. The Sui Dynasty continued the system of land equalization policy and the area of plowland and the agricultural production were greatly increased in a short 20 years. The grains and cloths collected by the Sui Dynasty were stacked up in the government's storehouse. The economy in the feudal society had been prosperous for that period of time, which was the result of adapting to the people's will.

---

① Xia Jie is the last emperor of the Xia Dynasty and he is a notorious tyrant in Chinese history.

② "Nine squares" refers to the system of state-owned land of ancient Chinese society. It appeared in the Shang Dynasty.

③ Emperor Xiaowen is the sixth emperor of the Northern and Southern Dynasties.

From 1840 to the next one hundred years, all sorts of developments in China fell far behind compared with many other countries. China was faced with difficulties from both home and abroad due to secluding the country from the outside world and not adapting to the people's will inside the country and not adapting to the trend of world's development. The reason why Communist Party of China can grow stronger and bigger, leading the whole nation to gain the victory of the War of Liberation is because Communist Party of China has adapted to the trend of the times, the popular sentiment, and the reality of China.

In all, from the beginning of the world to The Three Emperors and Fiver Sovereigns (in ancient China), up to now, the ups and downs of every dynasty and period, allude to whether the governing sytem adapted to the national condition, the popular sentiment, and the objective reality or not. It is the historical situation of China, so as for other countries all over the world: adapting, survive, or die out.

(4) We must also follow the adaptation laws in dealing with the international affairs. From civilized countries to uncultured nations, from empire to weak nations, the nation's development has proved that persons who disobey the law and violate people's will are sure to fail finally, no matter how strong they looked like. The proposals which are adapting to the people's will are sure to be more vital. The Five Principles of Peaceful Coexistence, which are used by China to deal with international affairs, have adapted to the people's will in the world; therefore, they have made great contribution to the world affairs.

Premier Zhou Enlai put forward the Five Principles of Peaceful Coexistence, including mutual non-aggression, mutual non-interference in each others' internal affairs, equality and mutual benefit, mutual resprct for territorial intrgrity and sovereignty, peacefully coexistence, which was not only confirmed in all the treaties, communiqués, declarations, bilateral relationship statements with other nations, but also have been quoted and reaffirmed in important international conferences and documents until now. The nucleus thinking of Five Principles of Peaceful Coexistence is adaptation, that is, every country should adapt to other countries' reality, interests and people's will.

All in all, from celestial body to galaxies, everything in the universe is in adaptation; all are in adaptation from nature to social reality; all are in adaptation at all times and all over the world; from Lao Zi's *Tao teh king*① (also *Tao Te Ching*) to platonic ideal country, every right thought has adapted to the objective reality; from Dante's *Holy Song* to Gorky's *Mother*, every masterpiece has been filled with the wisdom of adaptation. The ups and downs of dynasties prove that thing that adapts to the public's will can be prosperous, that does not adapt to the public's will would die. From looking into the truth to analyzing the fallacies, all can prove the difference between adaptation and non-adaptation. Adaptation is truth; adaptation is science; adaptation is the rule and it is the natural law. Non-adaptation is like a meteor. Adaptation can exist forever.

7.2 Principles of adaptation

Things in the world can not be separated from adaptation if they want to be good. Adaptation is the wish of people, and only the action of adaptation can get a smoother achievement. People all treat adaptation as the first principle of life in society. One should adapt to the parents; one should adapt to his children; one should adapt to the leader's arrangements; one should adapt to love the masses; one should adapt to be loyal to the country; one should adapt to respond to public opinion; one should adapt to conform to the objective laws; one should adapt to conform to the laws of nature. The developments of all things have their own rules, as the saying goes: "God does not change his routine, the land will not change the rules." The law is immutable and relentless. Adapting to the law, all will be prosperous; Not-adapting to the law, all will die out. Adapting to the law can make us understand the law better and find the problem, and then we can take the correct steps and make the best use of it. British philosopher Francis Bacon said: "We should only adapt to the nature in order to manage the nature." Lao Zi said: "Man adapts to the environment; the environment adapts to the heaven; the heaven adapts to Tao and Tao adapts to the nature." Here, Tao refers to the laws.

The theory of adaptation has three aspects: first, the correct understanding of things, including the law of things, social situations, public opinions, the major issues, etc. Second, it means to adapt to heaven, adapt to the envi-

① *Tao teh king*, known as the king of all the classics, is one of the greatest masterpieces in Chinese history.

ronment, adapt to the law, adapt to society, adapt to the fact, etc. Third, on the basis of adaptation, and only on the basis of adaptation, can we find the inherent laws of the development of things and the cause of the problem, reduce drag and friction, make the best use of everything, take measures and achieve the aim. Understanding of the law is the premise to adapt to the law. Following the laws is the practice of adapting to the law. Adaptation is a better way to understand the law. Adaptation is the premise, the means and the method. The ultimate goal is to achieve the aim by guiding and using it. The purpose of adaptation is to understand the law and make use of it. Adaptation is a soft power like that of water and it is also one of the most powerful forces. Non-adaptation will cause the opposite power. Although the opposite power doesn't burst out, it still exists. Such opposite forces would threat and be harmful to us. Adaptation does not exclude the struggle, violence and revolution. The progressive struggle, violence and revolution keep pace with the historical trend. Although the specific struggle is not the adaptation to the reality of the day, it is the adaptation to the future in a larger sense. The small non-adaptation is just for the larger adaptation in the future.

If adaptation needs an order, it should be nature first and society second, public interest first and self-interest second, large adaptation first and small adaptation second, major adaptation first and minor adaptation second, the urgent adaptation first and the less urgent adaptation second.

In order to develop something in the right direction, it is necessary to adapt to its internal law, adapt to the rational and main aspects.

Adaptation is obedience, acceptance, compliance, support and performance. Adaptation is a conscious and active action. It is not easy to achieve adaptation. If you know the importance of adaptation, it does not mean that you can achieve it. Adaptation needs endeavor and skills, including kindness, justice, leading spirit, capability of bearing insult, tolerance, wisdom, the respectful heart, the comprehensive ability, controlling force, goals and directions. Adaptation can not be achieved at will; it also needs experience, toughening and elaboration.

There are some old sayings about adaptation in ancient times. For example, Lao Zi said that the law should adapt itself to the nature. Modern people also have sayings about adaptation, such as "adapting to the tide of history", "adapting to the nature", and "make full use of the circumstances". In this book, adaptation is taken as the highest standard for the speaking and action of human beings; adaptation is the generalization of all the laws, so the essence of all the laws is adaptation.

Adaptation is the manifestation of development and civilization. The reform is to transform the things that can not adapt to the reality and the laws. The harmonious society is to live together harmoniously and smoothly. Doing things according to the laws is adapting to the law. Adaptation has been proved to be correct by the history.

7.3 The application of adaptation principle

In general, we must adapt to the law if we want to accomplish something; if we want to do something bad then we can do it in opposite way or adapt to it on the surface. We are successful in breaking up the old patterns and establishing the new ones, gaining the revolutionary success, because we have adapted to the objective reality and the public opinions. The world is colorful, but if we want to be fine, we must adapt to the law. As for "adaptation", first, we should know the necessity of adaptation and why we have to adapt to the law; second, we should know what we should adapt to if we want to adapt to the law, then we should know the law first (if we want to adapt to the main affairs; then we should know the main affairs first); third, make sure how to achieve adaptation according to the specific circumstance. There are two aspects to adapt to: to the rule, the reason, the public opinion, the law, the main affairs, etc. There are two ways of adaptation: one is that we should not contradict with ourselves; the other is that we should not contradict with the others.

Here, I want to explain the application of adaptation from the following aspects.

(1) The world affairs. The world affairs perform prominently in social problems, which refer to imbalance social relationship and circumstance, affecting the common life of the social members, breaking the formal activity of society, hindering the harmonious development of society.

The salient social problems of the world are: the relationship problems between countries, human rights problem, terrorism problem, populace problem, environment problem, energy problem, urbanization, grain problem, employing problem, juvenile delinquency problem, aging problem, trade protection problem, and so on. And the only way to solve these problems is to adapt to the actual situation, adapt to the specific circumstance of every country, adapt to the inner rules of every problem.

In terms of solving the problem between countries and nations, we should perform the principle of adaptation, and then we can solve it smoothly. For example, for the problem about invasion and being invaded, to invaders, first, they should admit that military invasion is dehumanization and unforgivable. This stain can not be eliminated forever just like man's crime. The invaders should take the responsibility for the harm and future trouble to the people in other countries. This is the rule for humanitarianism and they should adapt to it. Second, they should admit the crime and apologize to the victims. Third, they should take actions and make up for the victims, make more contributions to the whole human beings, especially on the part of protecting the world's peace, proving thoroughly to the world their determinations to rectify their errors and to avoid the repetition of the historical tragedy. For the invaded nations and the people in the world, they should supervise the invading countries by listing their words and observe their actions. This is a rule of adaptation and only after we adapt ourselves to each other then can we find the way to peaceful coexistence, mutual benefit and achieve common progress.

(2) Economic development. What we see is the surface, there are regular laws and reasons on the deep level. The economy law is the necessary link between economic phenomena. It manifests the essence of social economic development progress.

Economy laws are produced and play a role on certain economic conditions. If these economic conditions disappear, then the corresponding laws will disappear from the history stage. The economic conditions here mainly refer to the social production relationships. What kind of social production relationships we have determines what kind of the economic laws we will have. But when the old social production relationships disappear, all kinds of economy laws will disappear and recede from the history background. At the same time, new economy laws will be produced on the ground of new productive relationships. For example, there must be the surplus value law and production anarchy rule, where there exist the capitalist productive relationships.

The economy laws can be divided into two types, namely special economy law and the mutual economy law. The special economy law manifests the special essence of the economy process and economy phenomenon in a social formation. For example, the surplus value law is the special economy law in capitalism society. The mutual economy law manifests the mutual essence of economic development process in all social formations, such as the law that the production relationships have to adapt to the productivity. Every social formation is distinguished from others by its own special economic laws; and they are connected with each other by the mutual economy laws in all social formations.

The economy laws have objectivity and expresses that its existence and functions will not be transferred according to people's subjective will. People can neither "create", "establish" economy laws, nor "exterminate", "reformat" economy laws, but can only adapt to them.

We have to adapt to the economy laws and handle affairs according to the laws. The adaptation is a premise, and we should not twist it when there is a derivation, or we will be defeated by the laws. There are laws to follow when correcting deviations. The adaptation can know the laws well and discover the problems so as to be smoother and more active. Adaptation is for the sake of the more effective exploitation of economy laws and can guide things according to the economy laws. In fine, when we are using the economy laws, we should follow the principle of adaptation.

(3) Social activities. In daily life, adaptation brings you success and non-adaptation will bring you failure.

The functions of the government. The operation mechanism and management behavior of government agencies ought to adapt to the development demands of society, economy, culture, and public opinion, and play a leading role. The non-adaptation behavior of the government is sure to fail. For example, some government sections have ordered to move the freight cargo industry engaged in freight agent to the suburbs. Since the distribution market does not come into being there, and the freight transportation market must rely on it, eventually the government had to change their plan after several years of desolate running. It is just like climbing a tree to look for fish. Some sections has built a wholesale market in the improper area and as a result it is proved to be in vain. Some sections have taken over a lot of farmers' farmland for foreign trade and investment and the result is that the land have been wasted for several years. The failure of these actions is all because they didn't adapt to the inherent law of things and public opinions.

Industrial enterprises. The development of enterprises has to adapt to the need of domestic and international markets, national policies and consumers.

From the aspect of production, enterprises should continuously raise the scientific and technological level of their equipment; from the aspect of products, they should constantly make innovations according to market demands; from the aspect of after-sale service, they should keep on developing themselves towards humanization, taking people as their first priority; from the aspect of management, they should gradually decrease their cost; from the aspect of staff members, they should continuously improve the professional skills and ideological qualities of their staff members, and strengthen their sense of togetherness. All above are regular patterns and rationality which have to be adapted to.

Family education. There exist many misunderstandings of family education. Some children do not have a gift for arts and have no interest in musical instruments, while their parents force them to learn to play the piano and regardless of their wills. Parents are children's teachers, generals and talent scouts, not just commanders. Parents must adapt to their children's special natural instincts, and direct their children towards truth on the basis of adaptation. Parents should not beat up on children, nor threaten children with physical punishment; they can't be too harsh to their children, making children astray; they can't spoil children because it may lead children to choose a wrong path.

It is often the case that some parents with some knowledge are so confident of themselves that they consider themselves always right and like to impose their views on their children, which impedes the children's development and brings them strong resentment. Those children would never make any achievements. Some other poor educated parents who don't have the strong view believe that their children are better than them and give more free room to children. As a result, their children develop themselves naturally and healthily and it stands to reason that they make achievements.

(4) Conducting oneself in society. Likewise, the social interactions and the way people live and work with others can't do without adaptation. The natural temperament of water is to flow downward, running into the sea regardless of all difficulties; man's natural temperament is to develop upward, ceaselessly striving for a better life. A person's best life is of harmony and happiness. And in order to construct this kind of living conditions, people have to adapt to the world.

Employees and employers. The relationship between employees and employers is a matter of adaptation. Employees have to adapt to the employers from the perspective of the intentions of leaders, except matters of principles. Likewise, leaders have to adapt to the overall demand of benefits of all employees and all their public opinions. Otherwise, employers will lose their charisma. Not adapting to others will bring forth opposing forces. The relationships between superior and subordinate should be: the superior is an outline, the subordinate is a lucid exposition; the superior is a needle and the subordinates are threads, and the needle can direct thousands of threads to everywhere. If the superior has a right direction, the subordinates should follow him closely. If the superior is false, the subordinate can't be true; if the superior is true, the subordinate can not be false.

Husband and wife. Husband and wife should especially adapt to each other. As an old saying goes, "A good Jack makes a good Jill." For example, if your spouse likes to be flattered more than anything else, you should adapt to him or her to make your spouse happy. Otherwise, if you care other aspects instead of adapting to him or her, you'll gain just the opposite to what you wish. Adaptation to each other is the manifestation of true love.

Keep good mood. If we meet anything that makes us unhappy, we should first be aware of the followings: life is a process; life is a large stage where everyone is playing different roles; everyone has their own problems; this is what a world should be. After knowing these regulations and conventions, we should calmly adapt to regulations, rationality, sensibilities, and main affairs, and solve all problems in an orderly way. If we want to have a good mood permanently, we have to be kind, gentle and tolerant. Kind means to be kind-hearted and be kind to people. Gentle means to be polite, peaceful, and harmonious. Tolerant means to tolerate others and ourselves. One's occasional mistakes can be tolerated by himself. One should not blame himself constantly or over and over again. This is the truth of adaptation.

Adapt to different circumstances. People are born with different kinds of abilities. That's why they have to be engaged in the work which suits their abilities. But many people don't have the opportunity to do what they hope to. If we can make the best use of the circumstances, things may be successful. Even though a man is very frustrated, he should also adapt to the situation because not all efforts can bring good results and not all pays can be returned. This is the law and the reason of human beings. Everything has a law to follow, which we can't control. "A watched pot never boils." The more we want, the less we get. It's better for us to adapt to all. "A loss may turn out to be a gain."

Adapt to dialogue skills. Dialogue is the exchange of the soul and the most direct and perfect way of mutual understanding. First, we should hold a calm and irreverent attitude; this is the foundation of applying with the dignity of people and the foundation of good conversations. Second, we should listen to other people seriously, understand their thoughts and intentions, which is the inherent requirement of a communication. Third, adapt to others' words and guide their speaking on the basis of adaptation. If others' words make you feel surprised or angry when you talk with them, you should not give tit for tat and have a big quarrel with them. You ought to slow down and stop and think about what the reason is and whether what the other said is reasonable. At last, you can find the deep reasons behind the problem. Then you may shoot the arrow at the target and make remedy for the case so as to find the solution for the problem.

Adapt to your mind and body. Take it easy and rest if you're tired and go to bed if you're sleepy. Adapt to your own biological mechanism. For example, if one walks too much, his legs may be broken in a long strenuous march because his body cannot have a rest and complement. Heart prefers to be pure and quiet. "Fair and soft go far in a day." No peaceful heart, no gains. Living in the impulsive world, you shouldn't always force yourself to cater others; sometimes just follow your heart.

The thinking should be adaptable. It is easy for human brain to achieve deep thought if it adapts to the thinking. Sometimes you may find the more you don't want you to think about the things, the more you think about them. At that time, you'd better think from shifting perspectives. For example, when someone does something for you in return indeed, but if you haven't made allowance for the wishes of others and adapted to the trend, it is easy to cause resentment and hate.

Adapting to others is satisfying yourself. For example, if others hope you could follow and help them to attain their aims (without compromising on fundamentals), they will be satisfied if you adapt to them. It is the best if you adapt to them and you are also satisfied. If they need your help, but you give a hand with an insincere attitude, or rub them in the wrong way or in an ill time, we say this is like beating a dead horse. Of course you cannot compromise on fundamentals. Take another example, do not do to others what you would not have them do to you, because everyone has his own ideas. If you force others to accept your ideas, they will be ungrateful and even have some misunderstandings. Therefore, go easy on doing good deeds.

Master the overall situation. If you can put yourself in others' place, thinking things over, considering things carefully, inducing step by step, you will be unbeatable. Great statesmen and strategists have a lot in common: they all have a subtle insight into human nature; they can master the overall situation; they can adapt to the laws and make the best use of the circumstances. All things were related to each other. If you do something from the very beginning, you should think it over and make general evaluation and step-by-step implementation. If you take over a job in the middle, you should learn the reason, the origin and the former state in order to have a better connection. If you take over a job at the end, you should persist in overall consideration, properly handle the aftermath of the work and do a good job to link the future. This is adapting to the developing rules effectively.

Fighting against each other makes enemies, but adapting to others makes intimate friends. Most people have a stubborn temper, and there are always some ways and methods to defeat others. It is easy to do that. However, the result might be that both parties behave badly. If they adapt to each other and both parties turn out to adapt to each other, the outcome will be the manifestation of many highlights and advantages which have never been found before. In the end, both of them will think that the other side are good persons and will become close friends.

Take actions when the right time comes. People can't live without competition. Life is bitter and short; the key opportunities are few in one's life, especially in the market economy era and under the circumstance of cut-throat competition. Therefore, one has to compete fairly, struggling for his part, because this kind of fighting is also adapting to reasons. Of course, the struggling has to be appropriate; otherwise, it is not adaptation. One has to adapt to the brevity characteristic of human life, adapt to the human nature of contribution and the rationality of upholding justice and striving for the benefits for most people, adapt to the conscience of doing things well and working well. Absolute obedience, saving the situation by making unnecessary concessions, the meekly are not adaptations. Actually, they are against with the spirit of adaptation.

(5) The law of adaptation is widely accepted.

1) Adapt to the past, present and future. In the long river of time, there is only past, present and future. Adapt to the history, because history has already passed, which is not likely to go back again. What we can do is only to adapt to the fact, to sum up experiences and draw lessons from it. Adapt to the present: the present is what it is now. We have to adapt to the surroundings. Adapt to the future: we can predict the development trend of the future based on the circumstance at present and adapt to the conditions according to the development law of affairs and adapt to the lately-appeared predominant condition in the development.

2) Adapt to the home inside and outside, domestic and abroad. The activity space of people includes home both inside and outside, both domestic and abroad. All these places can apply the law of adaptation. The positive significance of adaptation is extremely obvious to Chinese people, because in Chinese people's eyes, "face" is more important than anything. But the westerners' cognition is different, they often like to speak directly and do what they like in the heart, and they are not afraid of conflicts, which they think may lead to consensus. In the west, people have to adapt to other people's nature and not copy the way of Orientals. This is the applicability and flexibility of "adaptation".

3) Adapt to the process before the event, in the event and after the event. Everything has its cause and effect. After the event, whether it is good or bad, the result can't be changed, but only to adapt, which is the only and best choice. According to the specific conditions of the fact, we should choose whether to accept it or learn a lesson from it. For example, in the family affairs, if the wife does something without discussing with her husband and the results are no so good. How to do with it? Adapting to the fact is not so bad; it is the best way to adapt to. What we can do is only to acknowledge it and adapt to it. We may reflect profoundly and draw lessons from it. Before we do things, we have to analyze the elements of things, find out the leading principle or elements, and adapt to the rules in order to make things develop smoothly. In the process of the event, we should focus on it, hold on to the major aspects, adapt to the inherent law, and accomplish it successfully.

Adaptation makes the world better. Everything will be better with adaption. The word adaptation can rule the world and the world can be prosperous with the word adaptation.

7.4 Short phrases about adaptation

(1) Phrases

For the world, adapt to the objective reality of the earth and the objective laws;

For the country, adapt to the specific conditions in each country;

For human beings, adapt to the human nature, human oriented spirit and human's benefits first;

For the motherland, adapt to the benefits of the motherland and nation;

For the work unit, adapt to one's own development;

For things, adapt to its inherent laws;

For people, adapt to his/her desire;

For the children, adapt to his/her natural instinct;

For the old man, adapt to filial piety to them;

For the couple, adapt to their love;

For the leader, adapt to obedience,

For the crowds, adapt to the public opinion;

For colleagues, adapt to cooperation;

For relatives, adapt to personal loyalty;

For neighbors, adapt to harmony.

(2) Folk proverbs

Let the nature take its course and guide in accordance with the prevailing trend.

Whatever happens, happens.

Forced love does not last.

Compelling doesn't make a couple.

One remains calm in face of fame or humiliation and holds a good mentality.

Not pleased by external gains and not saddened by personal losses.

A contented mind is a perpetual feast.

Be able to adapt oneself to different circumstances.

Make your life within your ability, you won't feel pain.

A man of great talents will surely be given important assignment.

Sometimes a foot may prove short while an inch may prove long.

What the lion cannot manage to do, the fox can.

A person was born into the world occasionally, which was his adapting to the times; a person was dead, which was also his adapting to the times. He is taking delight in his times and adapting to the natural laws. Neither sadness nor happiness can occupy his innermost being.

Man adapts to the environment; the environment adapts to the heaven; the heaven adapts to Tao and Tao adapts to the nature.

7.5 The solutions to "adaptation" problems

(1) Isn't it a word "adaptation"? Just adapt to it.

To adapt is not an easy thing to do. It is not that when we realize to adapt is right, then we can adapt to everything. Adaptation needs essentials of basic training, experience and toughening which include heart of politeness, heart of kindness, the ability to understand things or persons and the controlling consciousness of adaptation.

(2) Can we adapt to the enemy?

Of course we can. Adaptation is apparently conforming rather than yielding. It is adapting to the physical truth and taking effective measures. Justice is apt to defeat injustice. However, we should adapt to the enemy for national and mankind's benefits.

(3) How to conduct adaptation as things have various aspects?

Adaptation does not mean to adapt to all aspects but the inner laws and main goals.

(4) Leaders are used to saying "listen to me and do it according to my train of thought." Should we adapt to this?

This is a problem of adapting to the leaders or to the reality and the law. If the leader's thought is the law, it is right to adapt to; if what the leader thought is not the law or even against the law, we should not adapt to it. We should adapt to the law and adapt to the leader superficially and strategically in order to make him change his thinking. When nobody is clear about whether the leader's thought is right or wrong, we might adapt to what the leader said first and make it clear in the process of practice.

(5) How to deal with the situation where I want to adapt but the other side doesn't?

Everyone should follow the laws of adaptation. Adaptation will prosper and non-adaptation will fail. If you adapt to it first, it may become easier for the other side to adapt.

(6) Can the word adaptation rule the world? If I adapt, would the world be peaceful?

The subject and the object of adaptation over not on the same level. One subjcct's adaptation can't make much difference. For example, an ordinary Chinese farmer's adaptation or non-adaptation to the behavior of the President of the United States will have little effect. But if all Chinese people adapt to or do not adapt to him, it will have a great influence.

(7) Is there a someone who dislikes adaptation?

Everyone hopes adaptation. Even the thugs also threaten: "adapting to me, you will thrive; non-adapting to me, you will perish." Non-adaptation to the others is often driven by one's character and benefits.

(8) Someone serves the time and flatters others shamelessly. Is it a fact that he lives quite well?

The people who adapt to wrong path may feel good at a time. However, though if the thief's non-adaptation to the law made him rich, we can't think stealing is good and follow him. The thief will pay the price sooner or later for his non-adaptation. Adaptation is an irresistible trend, an orientation of the law. It will guide society into justice and establish healthy atmosphere.

(9) In fact, most people are adapting.

In actual life, people are basically in the trend of adaptation, and consciously or unconsciously agree with this truth. People on different level will adapt to different things on that level. People with vulgar tastes adapt to flattering others shamelessly; honest people adapt to logics and noble people adapt to major principles. In the unjust society, people with vulgar tastes may temporarily succeed, but they would be punished. The degree of justice of a society can be seen from the situation that people adapt to things. A society where people always flatter others shamelessly and the authorities are always right is an autocratic society. A society where there are a lot of robberies and stealing is an unjust society.

(10) What is the inspection standard of adaptation?

The standards of whether it adapts to things and whether it adapts in a right or wrong way are: whether the solutions to the specific questions can satisfy all the related aspects; whether it perfects the long-term substantial development of things and whether it makes progress along the healthy tracks.

As a matter of fact, man's nature is to adapt. They talk in dialect, eat what they are used to eating, adapt to the living environment generation after generation. Human beings adapt to their lives from cradle to the grave. Once they couldn't adapt, there would be some bad changes. When we trace something to its source for the reason, we may find that there's fame, gain and emotion in it. The essence is the innermost selfishness. And we have to adapt eventually. Only when we are able to adapt to things, expert in things and be happy with things, can life bring us good luck and success.

For us, there's nothing greater than everything adapts well, and all the theories aim at guiding people to live smoothly. All people's actions aim at adapting well, and all people are pursuing adaptation the whole life. Only when we adapt to the heaven, earth, society, the environment and ourselves, can we be happy. Human life should be based on kindness; human thought should be quick; human actions should be artistic; human figure should be beautiful; human will should be smooth. Kindness is the first among all the virtues; filial piety is the first among all the kindnesses; adaptation is the first among all the filial piety.

Chapter 8 Balance Theory

Balance exists everywhere and it is very easy to understand; If we really master its essentials, we can benefit a lot from it.

Balance theory exists widely in the natural world, in human society as well as in people's thoughts. We can use the theory to explain everything because it has universally instructional meanings and general practicality.

8.1 The balance theory

There are only two kinds of things in this world: the balanced and the unbalanced. All things in the world exist in the form of corresponding unity, in which the unity of opposites is included. Balance means that two or more corresponding aspects of a thing are reciprocal and complementary in quantity and quality; unbalance means that those aspects are not reciprocal and complementary in quantity and quality. Whether they are balanced or not lies in the ratio of quantity to quality. Balance has relative stability while unbalance has a strong tendency to develop further. Balance always ends up with unbalance and vice versa. Anything has a balanced relationship with things around, and the relative aspects of themselves have a balanced relationships, too. Therefore, you can find balance anywhere and anytime. The old balance is lost and then we have a new one, which makes things come into being, change and develop. Any development process, in fact, is a process of the continuous replacement of balance, the cause of which is the interacted unbalanced factors. Both balance and unbalance come into being in two ways: one is in the natural ways and the other is promoted to be. The former one always happens in the natural world and the other always happens in human society, so the progress of society sometime needs to be promoted. Balance embraces two qualities, namely coincidence and multi-dimension and under different situations; their developments have two qualities as multanimity and diversity. All things in this world exist, generally speaking, in a relatively balanced state but there are unbalanced factors everythere. Balance is fundamental, relative; and unbalance is universal, absolute.

8.2 The existence of balance

Balance generally exists in the natural world, in human society and in human's thoughts.

(1) Balance in the natural world

The existence of the universe is a mixture of the balanced and unbalanced matters. Everything changes from balanced state to unbalanced state, and then from unbalanced to balanced, which is a circular movement.

The earth goes around the sun. The movement, in fact, is a movement to seek a balance under the gravity among celestial bodies. Suppose there is a balance of something somewhere, we can say the balanced things are in (relative) stillness.

Take a building as an example. All the buildings, from the perspective of force science, are in balance under balanced forces. This is very clear and easy to understand.

Take explosion as another example. Explosion happens when the energy inside outweighs the energy outside (a huge unbalance) and explosion makes changes in its own nature. This example shows that unbalance can cause big or qualitative changes.

(2) Balance in human society

Human society is made up of many kinds of balanced and unbalanced matters or phenomena, such as the balance or unbalance in income and expenses, in production and marketing, in the ecological system, in aggregate demand and supply, etc. Specific examples are as follows.

Productive force and productive relations. The two always develop from balance (adaptation to each other) to unbalance alternately, thus pushing the society forward from lower class to higher one.

Deposit. If depositors deposit their money into a bank and the staff in the bank behaves well by providing good service, then in return the depositors would always come to the bank to deposit their money. In this way, the bank and the clients are in balance. Once the staff behaves badly and provides bad service, the clients would not get what they want and feel terrible. Consequently, the balance would break and the depositors would find another bank.

Employees and their lives after retirement. An employee spends most of his life working in an enterprise and offers a lot. If the enterprise can take the employees' later life into consideration and solve their troubles back at home, the employees would feel safe about their future; otherwise there will be unbalance between employees and enterprises.

Idealized ideas. People, especially children, like romantic things, and things about the brighter side of the society, so they have a fixed pattern in mind on romance, beauty and idealized things. They always expect too much. However, those are very far from the reality and even impossible in real life. Thus, a great unbalance appears. What is worse, those people live a less romantic and beautiful life than they could.

Maslow's hierarchy of needs①. In this theory, Maslow divided people's needs into five aspects: physiological needs, safety needs, social needs, esteem and self-actualization. Each one of them is a balance which is between the real life and things needed. Giving in the wrong layer of the five needs doesn't work or works badly and sometimes you must give a lot more at the wrong layer to get a balance. For example, some woman lives in the second layer, safety needs, and what she needs now is clothes to resist coldness. If you give her flowers, she will not be happy and the balance can not be reached then.

Hardworking. Hardworking and achievements are in direct ratio. Although the achievements not only include the achievements but also the acknowledgment, appreciation and gains in other aspects.

Mines. There are places in which coal mines and gold ore abound. These places have a lot of natural advantages over others. However, if they don't protect the environment and prevent soil erosion, they will have lots of future troubles. Therefore, where there are advantages, there are disadvantages brought by advantages. And the bigger the advantage is, the bigger the potential risks are. This is a homologous balance in opposite direction. The potential dangers, if not handled wisely, will be no less greater than the advantages.

(3) Balance in human's thoughts. What people think, do, speak and see

---

① A theory of psychological health predicated on fulfilling innate human needs in priority, culminating in self-actualization.

are all affected, unconsciously or consciously, by balance. Balance in human bodies and thoughts can be explained by the balance between losing and getting. Physiological balance is on human body. Nutrient balance makes one healthy and nutrient unbalance makes one ill; balance between *yin* and *yang* brings health and unbalance brings weakness. Psychological balance is on thoughts. Human thoughts are affected, unconsciously or consciously by balance or unbalance.

A contented mind is a perpetual feast. A contented mind adapts its expectations (also proper) to the real gains; thus it makes a psychological balance.

The function of "Thank you". When you say "thank you" to other people, it means that you get help from others and appreciate that. In return, you give something to them as thanks and compliments. People give first (help) and then your appreciation, and in this way they make a balance. But if people give a lot, just saying "thank you" may not make a balance then.

"Justice has long arms." A Chinese saying goes like this: you can not avoid wetting your shoes if you always walk along the river. People's thoughts and actions are dynamic and inertial. If not controlled and revised wisely, they will go to extremes and can not be held back. If the thoughts are on bad things, the thoughts will be worse; if you do something bad, you will do it again and finally put yourself against the law and society. "Justice has long arms", because a balance is made between what people do and what their actions bring.

Lack of self-confidence. People who are lack of self-confidence always underestimate themselves and have no courage and high aim so they are less likely to make high achievements in the real life.

8.3 Classification of balance

Balance differs in thousands of ways. We can classify balance into lots of types. Generally speaking, we can classify balance into two types: one is symmetrical balance which means that aspects of a thing are symmetrical in quantity; the other one is complementary balance which means that changes of one aspect in quantity or quality may cause changes of same quantity or quality of the other aspect in the opposite direction but the same in quantity. The two aspects complement each other to reach a balance.

(1) Symmetrical balance

Symmetrical balance means the relevant aspects of a thing is symmetrical in quantity.

Example 1: The balance of a marriage

Two people passionately in love may be attracted by mutual glamour. They may be attached to each other for the advantages of their partner. And they reach a balance in their mind. Then they get married. In marriage, they begin to realize their partner's changes in his or her mind and in everyday life. They no longer live a free single life as before and begin to worry about the fuel, rice, cooking oil and salt—daily necessities; they also begin to face all the social relations with all kinds of people. They become busier. This calls for the adaptability, which is an ability to make balance. There are many couples who think about happiness more than troubles before they get married. So they can not adapt to the life after marriage.

After they get married, they have to face housework, which is not very easy, not to mention their babies. If any one of the couple fails to do his or her part, such as being lazy in doing housework or always making their partner do housework, there would be unbalance between them. As time goes by, the one who works more, for not being understood or lack of praise, would feel unbalanced. He or she then would do something to make a balance. If the partner acts right next time, the balance is made. If not, there will be dirty words or crazy actions. And the family is in disharmony and the previous unbalance is worse or even lost.

Taking the above into consideration, we can see that every couple should make a new balance after they get married. Because the balance made during their romantic relations is no longer secure; that is to say, things are different now and love should also make relevant changes. When we say "marriage is the grave of love", we mean marriage is the grave of the previous love balance, and precisely speaking, the previous balance is lost.

Is it right that one's promotion means the balance will be broken? Not necessarily. First, one's promotion may have a close relationship with the other one's help or coordination, and if so, he or she would be grateful to the other one, and the balance is still there. Second, the promotion may give rise to wicked ideas, which demonstrates the multanimity of balance and the complexity of things.

Human beings are emotional. Mood rises or falls with emotions and a pacific mood ensures a pacific life. Couples should keep their partner's shortcomings in mind to make him or her cool when he or she is overconfident; similarly, couples should keep their partner's merits in mind in case of their partner's losing confidence. This can keep the balance effectively. Couples make proper self-evaluation and feel good for their life, and in return, they live a happy life. So we should learn to control the balance to make our marriage happy.

Example 2: The balance of relations between parents-in-law and daughter-in-law

Parents-in-law, when getting old, sometimes may be child-like or chatter without stop... This is natural for elders. Daughters-in-law should be extremely considerate towards their parents-in-law and talk with them, striving to make a balance which suits the elder's mind. The self-respect of parents-in-law should be fulfilled, and then the two sides can make a balance. Parents-in-law should, in return, care about their daughters-in-law, put up with their small shortcomings and praise their merits, striving to make a balance which suits the young people's mind.

Example 3: Getting on well with colleagues

There are always people who can not get on well with their colleagues. They think they are wronged and hold grudge to their colleagues. They think they do well and right but no one approve them, so they feel unbalanced. Actually, if they think more about their shortcomings and their unsuitable behaviors, they will feel balanced then.

Example 4: Striving to get out of adversity

On the way to be knowledgeable and during the journey of life, many people will meet adversity, such as illness, poverty, being disappointed in a love affair or in their political career, but they respond differently. There are always people who complain a lot and blame everyone and everything except for themselves. This is because they do not make a balance in their minds between their expectations (high) and the reality (low). Under such circumstances, if they choose to be depressed and pessimistic, they would meet endless unhappiness and never make a balance; if they choose to strive, to work hard towards a new height, they would possibly make an achievement in their career, and then they will make a balance, which, in return, comforts themselves.

(2) Complementary balance

Complementary balance means that changes of one aspect in quantity or quality may cause changes of same quantity or quality of the other aspect in the opposite direction but the same in quantity. The two aspects complement each other to reach a balance. Generally, if A is a positive one and B is a negative one, A a surplus one and B a deficit one, then A complements B and B offsets A to make a complementary balance.

In complementary balance, the higher part complements the other part. This kind of things always develops from complementary balance to symmetrical balance, and then move towards a new complementary balance.

Example 1: The anti-Japanese war

The anti-Japanese war best illustrates this question. The two parts of the balance in anti-Japanese war are Chinese people and Japanese imperialists. There were three stages in this war: On the first stage, the Japanese made strategic offense and the Chinese made strategic defense, which showed that China was a weak, semi-feudal and semi-colonial country and Japan was a powerful imperialistic country; they were in an unequally complementary condition and China was controlled in a long period. On the second stage, the Japanese made strategic contraction and the Chinese made preparations for strategic attack. There is somewhat a balance. On the third stage, the Japanese made strategic retreat and the Chinese made strategic attack. A new complementary balance was made then.

Example 2: Bourgeois revolution

Bourgeois revolution happened in the feudal society when the bourgeoisie was still weak compared with the feudal class. This was a complementary balance; when the bourgeoisie became strong, they made a symmetrical balance; when the bourgeoisie got victory, they made a new complementary balance.

Example 3: Advantages and disadvantages

There is a saying that where there are advantages, there are disadvantages. There are three implied meanings in this sentence: when the advantages outweigh disadvantages, they make a complementary balance; when the advantages equal disadvantages, they make a symmetrical balance; when the disadvantages outweigh advantages, they make a complementary balance.

Example 4: Balance between losings and gains

When you get something, you also lose something. This is also a complementary balance. For example, if you speak too highly of a child (give him too much), he may bebehaves badly next time.

The differences between complementary balance and unbalance:

Characteristics of complementary balance

a. Corresponding parts are not equal in quality;

b. Corresponding parts complement each other;

c. They have stability and continuance;

d. Corresponding parts are in a corresponding whole (unity).

Characteristics of unbalance

a. Corresponding parts are not equal in quality;

b. Corresponding parts do not complement each other;

c. They have no stability and continuance, but are changing all the time;

d. They have a strong tendency towards balance;

e. Corresponding parts are in a corresponding whole (unity).

Here are some examples:

1) Mentality and objective performance (complementary balance)

If a man feels satisfied about his current situation, he will not make a good performance objectively. The balance here is between his subjective satisfaction and objective lack of good performance, in which we can say his subjective satiation complements his objective lack of good performance, or we can also say his efforts and performance make a balance. This could happen in a stable state.

2) Expectations, efforts and objective reality (unbalance)

If a man does not feel satisfied about his current situation, then his expectations and the objective reality is unbalanced. This kind of unbalance has a strong tendency towards balance, and it encourages the man to make great efforts. Then there will be a balance between the subjective and the objective.

3) The water in a river (unbalanced)

The upstream and downstream of a river are unbalanced, so the water has no balance and can not keep still.

(3) Coincidence and multi-dimension in a balance

Balance embraces two qualities is coincidence and multi-dimension; its development has two qualities: multanimity and diversity. We have talked about the coincidence and the followings are some examples on multi-dimension and diversity.

For a man who works very hard and makes excellent performance, there is a balance between his desire to be good and his objective efforts; however, if his performance is not appreciated by his boss or colleagues, he will feel unbalanced; if he is appreciated, there is a balance then and he will work harder.

This is a virtuous circle. Suppose there are only oral commendations and no relevant material rewards, then the man would live a needy life, and there will still be no balance. Only if he can make his life better with material rewards, there will be a balance. Human relationships are complex. If you do better than others in your job, there will be unbalance and conflicts between you and your colleagues. Then, your colleagues may work hard to catch up with you or set you a barrier to prevent you being better than them. If this happens, you may help them catch up with you in their work to make them feel balanced. And you will have a new balance. Of course there will be developments and more balances.

The very same thing may develop into many balances or unbalances, which shows the complexity of things. We should work scientifically, flexibly and comprehensively.

(4) The classifications of the aspects in balance

The classification of the aspects in balance is very complex in symmetrical balance and complementary balance. Generally, they can be classified as follows:

1) A balance between the subjective and the subjective (psychological balance). The two aspects in the balance are all subjective sense, such as the balance between one's expectations and his psychological satisfaction.

2) A balance between the objectives. This means the two aspects in the balance are all objective matters, such as the balance between social demand and supply.

3) A balance between the subjective and objective or between the objective and subjective, such as the balance between subjective efforts and objective effects.

4) A balance between the spirituals and materials or materials and spirituals, such as the balance between mental shackles and material benefits.

5) A balance between the abstract and the concrete or the concrete and the abstract, such as the balance between one's thoughts, ideas and his actions.

6) A balance between one thing and many things or many things and one thing, such as the balance between a man and his surrounding environment.

7) A balance between mechanical things, such as the balance beam on a point.

8) A balance between different aspects of one thing, such as the balance

between different parts of a building.

(5) Graph representation of balance

In order to make the balance theory more clear and directly, we will use graph to help you understand and use it.

1) Symmetrical balance. Theory of symmetrical balance can be shown with two crossed straight lines with their middle points fixed on the y-axis; the parts on two sides of y-axis are symmetrical (see Figure 8.1).

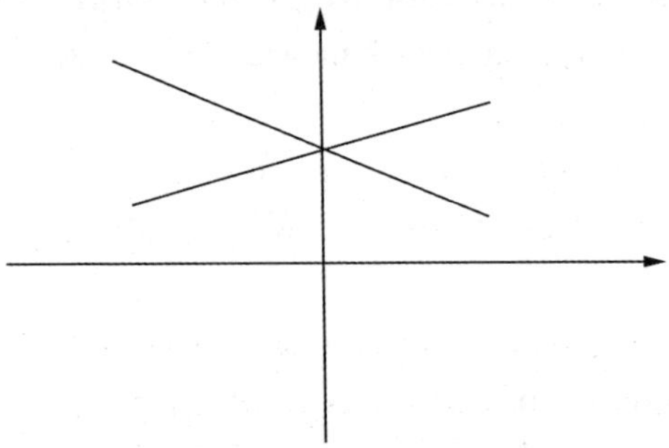

Figure 8.1 Symmetrical balance

2) Complementary balance. Theory of complementary balance can be shown as a straight line with its starting point on a 180° angle. The line moves freely and the two angles it forms is complementary (see Figure 8.2).

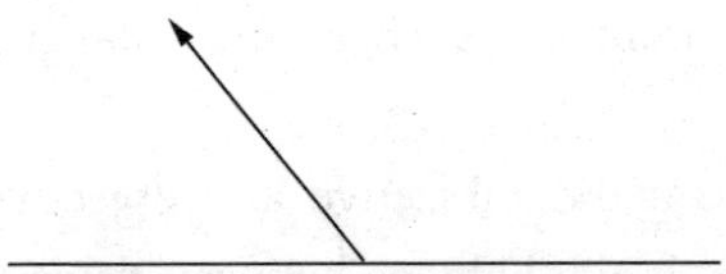

Figure 8.2 Complementary balance

8.4 Application of balance

Balance is a law for being and developing, and it exists in everything, no matter subjective or objective. If we can really understand and control it, use it to analyze and solve problems, we can certainly make bad things good, better, till best.

(1) Application of balance. If you want to put balance into practice, you should fully understand its regularity, analyze which balance is positive and

helpful and which is negative and harmful. We should maintain the former and break the latter to make new balance. We live in a complex society, but as long as we make good command of balance, we can save lots of efforts and make less mistakes.

We use some examples in symmetrical balance and complementary balance, but actually we use the regularity of balance to solve problems. We will explain it further in the following parts by using theory of balance.

1) Make a balance in your mind

Objective things exist in balance or unbalance, so does human's mind. Objective things are complicated and they do not change according to human's will; however, we can really make mental regulation happen. Why is human's mind unbalanced all the time? It is because humans don't feel satisfied easily, and they want more. They always complete with other people around him, among whom there must be someone who is better than them, so they feel unbalanced. There is a saying goes like this: if you keep comparing you with others, you have to die because there are always people who are better than you; if you keep comparing your goods with others' goods, you should throw them because there are always goods that are better than yours. You will certainly have weak points compared with others, so it is easy to find people who feel unbalanced around you. There are three ways to solve this. First, do not compare. Actually, every family has a skeleton in the cupboard. Everyone has his trouble, so we should just hold a positive attitude towards life instead of keeping comparing. Second, if we really want to compare with others, we should make it thoroughly. We should not compare our weak points with their strong points. Only in this way can we discover their misfortunes and make a balance in our minds. Third, learn to regulate your mind. The balance and unbalance in one's mind are closely related to the superior and the inferior. If unbalance is caused by the superior, you should keep that down; if it is caused by the inferior, you should keep it up. You can play yourself down by speaking highly of others, or you can build up yourself by playing others down. Here are two examples. If you think you are excellent and do a good job, but you do not get promotion, you would feel unbalanced. This is because of your sense of superiority. If you think more about your shortcomings and others' merits, you will feel better. If you get criticized and you can't accept it, you will feel unbalanced. The reason for this phenomenon is that you are in the sense of inferiority. If you think of your merits, which are better than the one who has criticized you, you will feel better and there will be a balance in your mind. Similarly, you can also think of their shortcomings to make a balance in your mind.

There was a story. In a holiday village, a girl was left in the tennis court after the children finished their tennis lessons. The staff counted the children, and found something wrong. The staff hurried to the court to find the girl back. The little girl felt she was forgotten, so she cried there.

So what should the girl's mother do then? Was she to comfort her daughter or to blame the staff? She did neither of them. Instead, the mother talked warmly and reasonably to her daughter: "it's OK now. But that sister feel nervous and sad because she couldn't find you. She didn't mean that, so go to kiss her to comfort her." The four-year-old girl stopped crying at once. She kissed that staff's cheek and said softly: "Don't be afraid, it's OK now." The staff cried with tears of joy.

So, you see, if the mother kept comforting the girl, she was telling her that she had just experienced a very horrible time. If she blamed the staff, she was telling the girl it was all others' faults and she was unforgivable. She would hold the idea that she had been treated unfairly and kept crying.

The mother solved the problem because she found the right balance and the right way to balance the child's mentality: the child cried because she thought she was nobody and abandoned. The balance aspect is to tell her that people around her care about her very much and they blame themselves for that mistake. Then, the child will have a sense of satisfaction and it's her turn to comfort others. She feels balanced again.

2) Interpersonal communication

It is easier to solve a problem when you find balance. For example, when you talk with someone, you should think about his intention and what kind of balance he wants to make. If you go against with their intention, there will be disagreements.

A potential car buyer picked a car in his mind and consulted the car seller. During their dialogue, he said excited: " My son was admitted into a key university, so we come here to buy a car to celebrate that." But the car seller just responded: "Oh, you should do that." Then he kept introducing the car. The buyer didn't listen at all and he just turned around and left, leaving the seller alone. After that, he phoned the buyer for the reason, and the buyer told him: " You didn't even care about what I was saying. I can buy a car at many places. Why should I buy it in your place?" So the reason is that the seller didn't make the buyer happy, which was a part of the balance, symmetrical balance.

One of a couple came back and said: " It's so hot today. I am dying!" If the one who was at home responded: " Come to the air conditioner and I'll fetch a cool drink for you", the one who felt hot would feel balanced and they would be OK. However, if the one who was at home responded: " You feel hot, who didn't then? Look at you, so spoiled", they would not make a balance and even make things worse and end up being unhappy.

If someone asks you for help and you help him with pleasure, you two make a balance and feel happy about it. If you hesitate or even refuse to offer help, you two will not make a balance and feel unhappy. Therefore, when others ask you for help and you offer help, they will appreciate you; if you do nothing when others need help but offer help when they don't need help, you will do it in vain even if you offer a lot. Giving a little bit to someone in need is better than giving a lot when others do not need.

People have balance in their mind or subconsciousness when they talk or do business. When they communicate, they always think ahead about what to do and how to do to make a balance. Then they take actions and make things go smoothly. Generally, if you give people the sense that they are important, you will get along well with others. Sense of self-importance is balanced with self-esteem. Of course, if you want to get along well with others, you should learn to give, to contribute, which are kind-hearted. To give means respect, understanding, forgiveness, consideration and relief; to give means cooperation, support and encouragement. For example, if your colleagues don't get well with you and you offer to support and care about them, you will soon make a balance in your relationship. This is the flexible application of balance theory.

3) Family education

Parents have great expectations for their children. If the parents think that it is great for a three-year-old child to count to ten, their child may not make great progress in a short time. Instead, if the parents think that it is not enough for a three-year-old child to know twenty characters, their child may learn thirty more characters. This is a balance between subjective efforts and objective effects for the children, and a complementary balance between subjectivity and objectivity. When one is complete, the other one will be incomplete, and vice versa.

We should praise more. People like to be praised and one praise is better than ten criticism. Think about what you would like your children to do in the future and find their merits in that field, and then you can lead them to that field. In some sense, parents determine their children's performance at school. The children will become what their parents want them to be. Proper praise makes children feel balanced with their efforts, and the children will work harder. Criticism and violence make children feel unbalanced. They will think they study for their parents and don't strive hard any longer.

Balance between parents and their children. There are disagreements and contradictions in many families. The reasons are as follows: a. Parents do not respect their children and they treat them unfairly, not as a complete person but a part of their fortune. They put their wills on their children and frequently harm their children's self-esteem. b. Parents are conservative and they can not adapt themselves to new things, which is against their children who are active and apt to receive new things. This is the generation gap and they can not bridge the gap. c. Parents are not well-educated and they don't know how to teach their children and can not understand them or convince them. All these make no balance at all. d. Parents and children are not responsible for each other. For example, the parents don't take care of their children and children don't do their part, either. e. The children behave badly and their actions and thoughts are not balanced with that of their parents'; parents only care for themselves which leaves bad effects on their children.

4) Gains and losses. There is a saying that where there are gains there are losses. When you get something somewhere, you will lose something else somewhere else or meet the potential losses. If you can not handle the problems properly, you will get corresponding losses. If you get something you do not deserve, the corresponding part will feel more unbalanced. They may go to extremes to make a balance and you may break the laws or go against social morality.

The ones who give more ask more; the ones who give less ask less. Receivers should also give, and the more he receives the more he will give, or else he will meet difficulties. This also applies to marriage. If one of the couple contributes more to the family (bring more fortune), he or she will want the other one to give something else to make up. If not, he or she will feel unbalanced. When there are something wrong they will behave terribly. Therefore, to receive is not necessarily good because you will lose something or owe something to the one who gives. If you do not pay back, there will be troubles because of the unbalance.

This is true for the givers and receivers. The ones who give you something good, such as money, gifts, labor, etc. , are the ones who are most likely to become estranged, because they want to be repaid to make a balance in their mind. If you forget their gifts, they may speak evil of you and feel unbalanced. So, you should remember to pay back. Similarly, if you do something good to others, they should also pay you back to make a balance between you and them.

Pains and gains. Pains and gains are corresponding parts for a balance. In some sense, if you want to get on well with others for long, you should not overdo something helpful to them. It is good to be natural because the more you give, the more you want back in your subconscious. If the other side does not realize that or forget that, there will be disagreements or contradictions. As time goes by, you will feel unbalanced in your mind and become estranged or break up with them. Just as the following sayings go: "The friendship between men of virtue is light like water"; "Even among brothers, accounts should be settled without ambiguity"; "To get the things you love more, you will give more efforts". On the contrary, if someone offers to help you a lot, you should remember to pay back, or you will displease him or her, and end up being unhappy.

When you get something, it will be easier for you to give or lose something. Usually, if someone gets something, he is likely to ignore the facts or go beyond himself or be cheated. Then risks appear. For example, the fortune-teller will flatter you before he tells your fortune; you will get profits or win when the first time you gamble. It will be easier for you to be trapped if you believe in the products on discount, free trial and testing before you buy, because you think you have got something free. Similarly, if you get a bonus of 500 *yuan*, you may go out to spend some of them to celebrate. But if the money is your salary, you would not do that because you take it as hard-won money and you will spend them on urgent things. The above happens because you get more when you get bonus, at least you feel so. That is beyond balance.

5) Spirit and material. The heroes at war are honorable and this is spiritual reward, whose corresponding part is material. If we can not ensure good

results for the both, neither will be ensured. Suppose there are only spiritual rewards as publicity and praise or models making reports and they are not materially ensured, the spiritual part will be lost, too. Of course, this will not leave good influence to the society; on the other hand, if the models always think about asking for material benefits, his spiritual part will be lost. So the government should give some material rewards, which is a very important part in the political and ideological work. If they live a well-off life, only spiritual rewards would make them feel happy and balanced. We should solve problems according to the reality.

Why are the heroes in the wars feel satisfied? Because there is a balance: in war times, their lives were as terrible as the others were and they would fight not only for others but also for themselves. If they didn't fight, they would suffer.

6) Misfortunes and fortunes. Misfortunes are also fortunes, for misfortunes make People ponder on their life. Their hopes to break the balance becomes stronger and their wisdom will be fully used. This makes them to change their current situation to make a new balance. They understand the difficulty to make their life better, so they treasure what they have had and this ensures a long time balance. Good luck is not only an excellent opportunity but also a fetter for success, for good luck makes people proud and lazy. People's wisdom stops working, and the previous balance will fall into a lower level to fit their laziness and pride. The saying that "one should be mindful of the potential dangers in time of peace" also tells us that.

7) Adversity and favorable circumstances. In adversity, one should be mindful of their advantages to avoid the lower balance between the depressed ideas and the adversity. He should strive with spirit to make a higher balance. Under favorable circumstances, one should think of their disadvantages and shortcomings to remind him of trying hard to maintain the balance.

8) Cherish your experience. Cherish your past, present and future, for the present is the best, the past is valuable and the future is hopeful and determined by the present. Cherish what you've had and lost, for what you have may be lost one day and what you've lost may bring you something. Cherish your success and failure, for they both are treasures. Cherish your experience and lessons, for they both are valuables. Cherish all that you experienced, for there will be an end to that. So, to cherish is to maintain and establish a better balance.

9) Pride leads to loss while modesty brings benefit. The one who talks big or blames others is always the one who fails his goal or the next to be blamed. This is because he is so proud that he thinks he is wiser than others. So, he doesn't even analyze how to solve the problem and avoid the same mistakes. He blames others because he aims too high which is against the real condition. He aims too high with lower abilities, and this is the complementary balance. If he learns lessons from others and think over the real conditions, he will make his mind tally with the conditions to avoid mistakes (symmetrical balance). Therefore, what misfortunes on others today may come to you tomorrow. There are endless examples.

If you want to get things done, you should underestimate your ability a little bit and show yourself at proper extent. People who say they are needy live a good life, whereas people who always speak highly of his wealth suffer their life. This is complementary balance.

10) Respond to criticism kindly. When solving problems in hostile situation, you should listen to other people's criticism and suggestions calmly and not argue. This can satisfy them to make them feel balanced, which makes the problems be solved easily.

11) Self-esteem maintenance. Dale Carnegie① said that if you follow a rule, you can make many friends; besides, you can avoid many disputes and find youself happiness. The rule is to make others feel their importance. If they think they are not important to you, they will leave you; if you make them feel they have many shortcomings, they will hate you. Therefore, avoid showing that you know their shortcomings, for people have self-esteem. When they feel they are important, they feel balanced; when others regard them as unimportant ones, they can not maintain their self-esteem and feel embarrassed, so they don't feel balanced at all.

12) Balance between loves. One of the young couple in love affairs always asks the other one to say "I love you" to him or her. It means that he or she has already fallen in love with the other one. He or she forces the other one to show the same passion as he or she does, and then he or she can make a balance. There is a risk: if he or she is not satisfied, love may turn into hate. You should show your love if you receive love from others, and that's

---

① Dale Carnegie(1888—1955), known as the father of modern adults education.

the way people make balance in love.

13) Shifts in a balance. If someone who are lower than you beg you and you do not give a hand, they will complain and you will be embarrassed. Then you become the one at a lower place: you may have to apologize to make them feel better (balanced). If you get mad at someone, such as your wife, but she doesn't talk back or respond, you are at a higher place than her. After that, however, you need to apologize and you shift your place with hers and you are at a lower place then, and then she will feel balanced. So, aspects in a balance keeps shifting and they always go to the opposite direction.

14) Doing business. Negotiations, in fact, are the process of making balances. You should make the unbalance into balance which can be accepted by both parts you will sometimes break the previous balance to make a new one.

A salesman of skin care products goes to a consumer to introduce his product, but the consumer says: "I don't want to buy that, please leave me alone." If you are the salesman, what should you do then? To go away or to convince? The answer is the latter. You should learn about the consumer's information on consuming, which may be one of the three aspects of a balance (balance here means his mentality for consuming). First, the consumer is using other brands; second, the consumer was cheated before, because the product has no effects or side effects; third, the consumer never used skin care products before. After getting all the information, you should help him or her to analyze, to compare and convince him or her to solve the problems. You break the previous balance and make a new one which is more helpful for the consumer.

Marketing does not mean sell products only. More importantly, what you should sell is your service, care, help and so on. Your sales volume is actually determined by all the above factors. Sometimes, consumers ask you for help, but their problems are not about your products but services which has nothing to do with your products. If you try your best to help, they will appreciate that and feel balanced. In return, they will believe your products and may become your potential consumers or fixed consumers. Your success lies in them.

15) Make wise use of balance. In real life, no matter when or where, you can make situations good for you if you make wise use of balance theory. If we compete with each other, my winning shows your failure. If I lift my position, you will become lower than me and we are not in a balance. For exam-

ple, two people make an appointment and the one who come ahead of the other gets the upper hand. Similarly, if I am more active than you in doing our own business, I will be the one who control our business. For example, I come earlier in a date and text message you, and then I will be the one who control this date. You wear decently before others, and you will be superior in some sense. In negotiations and competitions, you can make the situations advantageous for you. When you meet someone and you speak highly of him, the situation is also advantageous for you because you give and they receive (compliments). When someone asks for your help, you should keep their expectations lower, and whether you help them out or not at last he can easily make a balance in his mind.

People in a contradiction always have different mentality. If they change their mentality, the contradiction will be solved easily and they will feel better. When two people communicate with each other, there will be the one who have advantages and this is unbalance. But if the one with advantages still performs decently, the unbalance will be worse and the great figure appears.

When a man makes a mistake, he will, firstly, be afraid of being criticized and feel unbalanced before others. Then, he will try to make excuses to get rid of the criticism to make a balance with others. If he is criticized and lose face before others, he will defy and give many excuses to make a balance instead of thinking over his own mistakes. If he is not criticized even though he makes a mistake, he will make a new balance and maybe rethink his mistake to warn himself not to make that again. Therefore, to criticize is not always a good thing.

If a man deserves a position but he does not get it, he will not feel balanced. And maybe he will be more depressed because of others' acknowledgment.

Keeping your expectations down will be helpful for you to make a balanced mentality. For example, people will be unsatisfied if he gets less bonus because his company meets difficulties in operating. But if he is to be fired by this company and the manager tells him he can stay, he will be very happy, even though he only gets a small bonus or even no bonus.

In a word, human beings break their current balance and make new balance. They move forward from unbalance to balance, from balance to a new unbalance. Things develops into their opposites. The positive attitude should be like this: break the current balance to make a better one; make a better balance change into unbalance to push forward the developments of these things.

(2) Put balance theory into practice

This may include the following aspects:

a. Maintain the active and helpful balance;

b. Make the current unbalance into a new, better balance to make a good start;

c. Avoid negative balance;

d. Break old, negative balance into a new, positive one. Make that a tendency and finally we will have a new, better balance;

Friendship is precious and we should try hard to make it long-lasting. When a man is determined to be a successful man, he makes an unbalance as long as the determination is there. The unbalance is between his determination and the fact that he is not successful now. If he knows himself and makes a proper goal which he can reach if he tries hard, he can really reach his goal to make a new balance as long as he makes effort.

(3) The changeability of balance

The universe is complex, so the balance theory which show the law of universe is also complex, among which there is multiplicity and changeability.

Sometimes it is very difficult for people to find his own balance. This is determined by his emotions and positions. The saying "those closely involved cannot see clearly" also tells us this. Under such conditions, he can turn to his friends for help to make a proper balance.

As the proverb goes: "Harmony brings wealth." If you do good to others, they will do good to you in return. However, influenced by character, someone still do evil to others even he knows it will bring back evils. Consequently, he lost more. Therefore, we can say that nice character is a blessing and to some extent, character determines destiny.

You are nice to people, and that is what you give to people. They will be nice to you as a return. On the contrary, others do good to you, and you should remember to thank them in some way. If people who receive help do not do good to others in return, they will be looked down upon by others and what they got will disappear.

The proper balance will be twisted or even broken if people are different in educational level, social experience, moral qualities and living environment. Under such conditions, we will have balance or unbalance beyond our expectations.

Chapter 9 Structure Determines Everything

Structure is not mysterious, but its effect is enormous and incomparable. Every law in the world can all be contained in the structure.

There is a well-known story[①] in ancient China:

In the Warring States Period[②], King Wei of Qi State[③] was fond of horseback riding and archery. He often competed with others and could win nine times out of ten. One day, King Wei proposed to match against Tian Ji[④], the marshal of Qi State and bet on a thousand pieces of gold. Tian Ji readily agreed. Nevertheless, he was still wondering how he could win the race this time since he lost several races to the king before.

Returning home, Tian Ji told the thing to Sun Bin, a brave and astute strategist. Being an official in Wei State at the very beginning, Sun Bin flee from Wei State to Qi State to avoid being set up by the vicious and sinister Pang Juan who was blinded by greed. Having known Sun Bin was an honest person as well as a master of strategy, Tian Ji admired Sun sincerely. He let Sun stay in his home and treated him as the most respectable guest. With shared interests and mind, they became close to each other like brothers.

Sun Bin asked Tian Ji: "what are the rules of the game?"

"Each one will prepare three warhorses, which are to be classified into the first,

---

① Horseback Riding is a famous story about how to win in a competition by making best use of the advantages and bypass the disadvantages in ancient Chinese history.

② The period of the Warring States (475 B. C. —221 B. C.) commenced at a time when the numerous petty city-state kingdoms of the Spring and Autumn Period had been consolidated into seven major contenders and a few minor enclaves in ancient China.

③ King Wei of Qi (378 B. C. —320 B. C.) is the fourth king of Qi state during Warring States Period. His reign lasted from 356 B. C. to 320 B. C.

④ Tian Ji, birth and death unknown, is a famous general during the Warring States Period.

secondary, and third class. The first-class horse will compete with the first-class horse, the secondary-class with the secondary-class and the third-class with the third-class. Victory belongs to the one who wins the round. My horses were insufficient in strength and thus I lost every match to the king", replied Tian Ji.

Thinking for a while, Sun Bin said, "I am sure you'll win this time."

On the match day, Sun Bin said to Tian Ji, "You put the best bridle and saddle on the third-class horse, making it the best horse to compete with the king's best horse. Then let your first-class horse compete against the king's second best horse and secondary-class horse compete against the king's third-class horse."

Tian Ji get fully prepared according to Sun Bin's idea. Then, he urged horses to the court, waving his arms and bending the bow, racing against the king back and forth, contending for the victory. Tian Ji lost the first race, but won the second and the third races. As a result, he won the reward of a thousand pieces of gold.

The result was beyond King Wei's expectation, making him feel very weird. He asked Tian Ji how he could win this time. Tian Ji told the king truthfully about how Sun Bin helped him come up with the idea. Hearing the whole story, King Wei sang high praise of Sun's wisdom. Thereafter, King Wei boldly put Sun Bin in an important position and let Tian Ji and Sun Bin command the armed forces of Qi State.

The above story vividly tells us a philosophical reason: The table can be turned by transforming the structure.

Structure decides things' nature and laws. Knowing things is to know their structures. Innovation is to make beneficial changes to existent structures.

9.1 The definition and classification of structure

9.1.1 The definition of structure

Structure is the form of things' existence. Everything in the universe is made up of elements, whose mixture and arrangement form the structure. The relationship between elements and the outside world constitute the structure's contents.

9.1.2 The classification of structure

(1) In terms of complexity, structure can be divided into monolithic structure and group structure. The relations among several structures and the connections among structures make up structure group. Structure group and its relations with the outside world constitute the whole content of group struc-

ture. Structure can nearly be decomposed infinitely. The final result of decomposition shows: substance is made up of the most basic and simplest structural unit; structure can be overlaid endlessly, and the universe is made up by countless different structures. Every substance is the same at the premises of scrutinized dissection. There exists correspondence and similarity between each substance. Therefore, physical state is often used to explain the matter of thinking, for mind is substance too.

Group variety structure is the main form of things' existence and development. From the galaxy to microbe, all choose group variety as the form of its own development. The development of group variety is affected by three factors:

1) the positive or negative impacts outside the group variety;

2) the positive or negative impacts inside the group variety;

3) internal self-discipline.

Substance, particularly living plants and animals, exists and develops in the form of group structure. Only by integrating itself into its group system, can each individual define its identity and usefulness.

(2) In terms of category, structure can be divided into mental structure, social structure and natural structure in terms of category. All structures related to mind belong to mental structure, including thought, consciousness, etc. All structures concerned with society belong to social structure, including human society, animal groups, and so on. The natural world with the exception of fauna and plant population belong to natural structure.

The universe is formed by independent objects and their relations. Every monolithic structure has exterior relations. Thus, things are universally connected. The universe is a large structure where things are widespreadly connected with each other. The universe is boundless, with a huge variety of species. It is in order of time sequence longitudinally and is shaped by structure laterally. All the things in universe forms universe structure; all the things in natural world form natural structure; all the elements in human society form the structure of human society; every nation has its structure; everything has its structure.

9.2 Attributes and changing forms of structure

9.2.1 Structure attributes

Structure has the following attributes:

(1)The factors that make up structure are interdependent. One structure and relevant structures are interrelated, depending and acting on each other;

(2) Structures can be changed and transformed;

(3) The adjustment of structure is the result of the synthesis of internal factors and external ones;

(4) The independence of structure;

9.2.2 The changing forms of structure

Structure is changed through the following forms:

(1) Change according to the order of sequence;

(2) Change in accordance with planar structure;

(3) Change according to three-dimensional geometry;

(4) Change in accordance with quantity relative ratio.

Things rely on their adjuncts, which are mainly layer structure. Layer is the main form of structure. For example, building, air, earth crust, bedding, books, clothes, skin of animals and plants, furniture, idea, article, molecular structure, and so on. Numerous things basically exist in the form of layer.

Mechanism, the soul of the creation of structures, is a new structure in the mean time.

9.3 Structure determines everything

The universe is made up of of material. Material comes into being in the form of objects, which exist in the form of structure. Things are comprised of elements, whose combinations are structures. To change elements is to change structure as well as to change things' nature and laws of movement. Therefore, it is structure that determines things' nature and laws of movement. To know things equals to to know their structures.

Anything's nature is determined by the amount of elements (the amount among internal elements and between elements and the outside related things) and proportion and its arrangement (internal arrangement and its structural arrangement with the outside). Thus, amount and proportion are a form of structure too. For structure, there are the change of quantity and the combination of oriention dimension.

Substance is made up of proton, neutron, quark and even smaller particles. Even if we decompose any forms of substance in a microcosmic way, the material form and ingredient will remain the same.

The reason why the forms of substance are varied lies in the different structures. The different amounts of elements and arrangement structures determine things' different nature. To change a structure is to change its elements, quantity relation ratio and arrangement.

Structures are everywhere and constantly play decisive roles. Marx said, "There would be no railway today if we have to wait some individual capital to grow to such extents that have enough capita to build railway. Nevertheless, the matter of capital was easily solved by means of stock company." That demonstrates the power of changing structures. During the warring states period, the structure of multi-states determined that its social nature was division and annexation. After the structure of unification came into being, history has been changed tremendously. In the era of polarized world, two superpowers monopolized world affairs. Nowadays, multi-polarized development brings a new image to the world. As an immigrant country, America's multinational and multiracial social structure contributes to its unique feature of being multicultural and open-minded. The mechanism of an organization is in its own structure, which decides upon the rise and decline of the organization. If there are too many levels of administration in a country, they will lead to inefficiency of command and response. Instead, if levels of administration are arranged in a scientific way, efficiency will be improved accordingly. Combat and battle field lineup is a structure, and different structures bring in different results. For instance, the well-known battle of "Ma Su[①] lost Jie Ting Fort" in *The Romance of the Three Kingdoms* culminates with errors in battle field lineup. Wooden ox and horse invented by Zhuge Liang is a typical example of structure creating a miracle. The reason why the "Long Men Zhen", a famous battle field lineup becomes so mighty lies in its fabulous structure. Structure explains why there

---

① Ma Su (190—228) is an official and general of Shu-Han during Three Kingdoms Period.

exist omniform buildings. Substance's molecular structures determine its physical and chemical property. The nature of substance is determined by the number and arrangement of molecule, proton, quark or even smaller particles such as ethanol and methyl ether which share common molecule but not same structures. Their molecular composition and molecular weight are totally the same, but their molecular structures are different. Thus, their physical property and chemical property are rather different. Gene determines species, structures determine nature. Researches on brain testify that a person with high intelligence does not necessarily have large cranial capacity. Brain imaging techniques proves that differences may lie in the ways of cranial development. The structure of brain determines intelligence. Violent people who tend to act impulsively have inefficiencies in brain structures. A person's inborn qualities determine that he or she won't reach the heights outstanding persons have achieved even with lifelong practice. Due to the different body structures, beauties exhibit curvy stature in a delicate and beautiful way. In singing, from high, low, long or short song tune to mouth shape, standing posture, the way airstream flow through and the expression of emotions, all the above are structural forms. Every language has been formed by its standard structure. Different sentence structures would express different meanings, for example, the three phrases "All for the children", "for all the children" and "for the children's all". The meaning of each phrase changes accordingly due to the different order of words. Woodpecker could protect its brain from being harmed for its skull and brain structures: Its sensitive brain cell is surrounded by muscle, which makes woodpecker's brain serve as a perfect bumper. Owing to different structures of organism, chicken will never fly like eagles. One's handwriting can have his style by reason of structures. With its own elements and structures, a properly arranged passage can be called a good passage. Or else, it merely amounts to nothing. One's way and procedure of doing things are structures too; appropriate structures could ensure success rates. The framed formula of thinking is structure as well; changing undesirable thinking set would create miracles.

In the boundless universe, variously complicated things develop in many different shapes. All these phenomena have their roots in the variance and changes of structures.

9.4 Change the structure and develop innovatively

Innovation is to change structures, which means to make beneficial changes to what is already existent. Here changes mean changing structures. Objective existence is widespread. Thus, innovation is widespread and infinite. Changing structures often resort to reverse thinking or looking at things from new methods and new perspective. Therefore, to change structures is to change ways of thinking. Discovery, invention and innovation are all fulfilled by breaking the routines, so in some sense mistakes and errors become a way of making innovations.

Miracles can be made by changing structures. The changes of things are unpredictable; however, the root causes lie in their structures. A popular saying goes, "Nothing can be accomplished without circle norms." Circle is the manifestation of rules. Sometimes it is square, sometimes it is round and sometimes it is the combination of squareness and roundness. Actually, circle's nature is a structural form. Lots of things can not be done in a routine way; instead they will be done by making some changes to the exact situation. If we all have better development conditions, what we need to do is to change structures of combination.

Creativity lies in the people. There would be no vitality without innovation, so are family life and social affairs. We would be imbecile if we rigidly adhere to simple, basic and repeated works. To innovate requires thoughts and organization structure of innovation and the establishment of new mechanism. System and mechanism are the matters of structures. The structure of innovation will be set up once we reach the goal of civil participation. To give full play to people's innovation ability, a decent environment and innovation platform must be built. Incentive motivation and spiritual guidance are the only effective way to make it become a reality. People are inexhaustible sources of creativity. When all people's wisdom of innovation is utilized to the fullest, the country is undoubtedly at the apex of its innovation.

9.5 Case analysis

We have already listed some examples in the above passages. Here are some more examples.

Example 1: Material structure determines nature. While traditional copper cable allows 10 million messages transmitted all at once, optical cable could transmit 1,000 billion messages simultaneously. That disparity results from different materials. Essentially speaking, its changes are caused by different structure. Likewise, the sizes of electronic products with the same function become smaller and smaller due to the upgrades and improvements of materials.

Example 2: When Zhang Liang, a strategist of Emperor Gao Zu of Han Dynasty, was a disciple under Sir Huang Shi①, who sold scissors in the day and read books in the evening. Later, he found himself not fit for business. Being not good at bargaining with customers, he did not make good bargains though selling scissors took much of his energy and time. In addition, he felt short of time for reading. Thus, he just classified his scissors into three classes, the first-class, the second-class and the third-class. The price of the first-class scissors remained unchanged; the price of the second-class scissors was cut down by one penny and the price of the third-class scissors was cut down by two pennies. As a result, the amount of scissors Zhang Liang sold that day was twice as many as he normally did and gained twice as much as he normally did. The story is the source of the famous Chinese saying: "As Zhang Liang sells scissors, the quality of scissors is same even though their price is different."

Example 3: There's such a test: On a stormy evening, you are driving your car through a bus station, where there are three people waiting anxiously for bus. One of them is a dying old man, who needs to be sent to hospital immediately. Another one is a doctor who saved your life before. You even dream of repaying him in your dreams. The last one is your dream lover. If you miss her this time, you will probably be filled with remorse for the rest of your life. However, your car can only seat one more person, what are you going to do? You should first try to save the old dying man. You also want the doctor to be in the car since he once saved your life and this is a great opportunity to repay him. Then comes your dream lover, you may never meet such a girl who will touch your heart so deeply if you miss her. What's the ideal solution? Here it is: You hand the car keys to the doctor and let him send the old man to hospital. And you stay to accompany your dream lover, waiting for the bus. That is the magical result of structure adjustment.

---

① Sir Huang Shi (292 B.C.—195 B.C.) is a scholar of Taoism.

Example 4: Social structure determines life quality. As the proverb goes, "The unpleasant matters can occupy 8/10 or 9/10 of one's whole life." Why are we dissatisfied with our life? It is mainly owing to our life style and social structure. The root cause lies in selfish motives, which can never be satisfied. Dissatisfaction would result in discontent, vexation and unhappiness. From the social point view, selfish motives are characterized by private ownership, which serves as the source of evils. The avarice of family is a bottomless pit. Supposed we make some changes to social structure by getting rid of private ownership, the treasure of society can still meet the needs of all the people. Moreover, people will feel happier. Changing structure of life is an effective method to change people's unhappy mood at the present time. Of course, new structure will bring about new problems. Nevertheless, it is a matter of pros and cons and is a constant improving process.

Example 5: World structure is the deciding factor for our safety. World's political, economic and military structures are moving toward regionalization and federalization. The establishment or potential establishment of European Union, North Atlantic Treaty Organization, North American economic and military alliance, South American economic and military alliance, Middle East economic and military alliance, African economic and military alliance, European and Asian economic and military alliance mainly made up by Russia, China and India as well as South Asian American economic and military alliance would change the supremacy of the USA and make the world situation more stable and reasonable. This progress, if handed well, will make the world a more harmonious place to live in. However, there exist the potential hazards of big-scale conflicts between region and region, alliance and alliance.

Example 6: Aggregation creates effects and vastness yields power. "Effects" in "aggregation create effects" refer to the distinctive and bigger effects. That is to say, aggregation yields bigger effects. With the aggregation of population and resources, city's development speed is several times quicker than scattered village's speed. For the aggregation of students and teachers, university's efficiency of imparting knowledge can not be compared by old-style private school. With the combination of favorable policies, good soft and hard investment environment and investors, the development speed of special economic zone and economic development zone tops regions. Silicon Valley is comprised of high-tech enterprises; its profits top the whole world. Congregated by a large number of high-tech talents, patents acquired in California University in 2003 surpass India and China alone, while patents IBM acquired is as five times as the number of patents that India and China acquired. The major international activities such as the Olympic Games and the World Cup are held in every four years, leaving people filled with expectations. Their effects of aggregation far surpass the effects when they are held annually. On Children's Day in 2006, CCTV (China Central Television) launched a large-scale children program "Dream Paradise" in which all sorts of characters that impressed children deeply are included. Once broadcasted, the program received favorable responses from children as well as parents. The key of its success lies in that it gathered all characters familiar to children. The feeling of yearn creeps up on one when he has a short parting from his relatives. The long time parting would generate a great deal of longing, which tend to burst out when he reunites with his relatives. It is thus clear that aggregation yields effects and competitiveness. Aggregation itself is power and development. The effect of aggregation is one plus one exceeds two ($1+1>2$).

There are three basic conditions for the creation of accumulated effect. First, there is a suitable environment for accumulation, including soft and hard investment environment. Second, accumulative individuals have common objective requirements and the environment is constantly meeting their requirements. Third, the objects being accumulated together influence and stimulate each other, setting forth tremendous appeal and creation, forming the atmosphere of the swelling of same values.

The forms of accumulation are divided into two kinds: One is the expansion of points, which means the expansion of one point; the other one is the expansion of areas, which forms networks.

Vastness generates power. The sun is large and it has lots of energy, while the moon is smaller and its gravitation is smaller. The big ocean has large amounts of energy while the small pond has relatively small energy. A large country always has a large market and huge spending power (some countries may not have huge spending power at present). The big enterprise has enormous might and capital, and therefore it has big influence.

In the course of human society, a crowd utterly lacking cohesion would achieve nothing. Something can be achieved only when they work together. As the mass is organized together for a single shared goal, the power is mighty and irresistible like the momentum force of an avalanche. All these are owing to the power of structures.

9.6 The use of structure

(1) Structural analysis. Since structure is made up by elements, we should start with elements as long as we want to analyze structure or achieve something. Here comes a method of structural analysis, and I hope it will bring you some enlightenment.

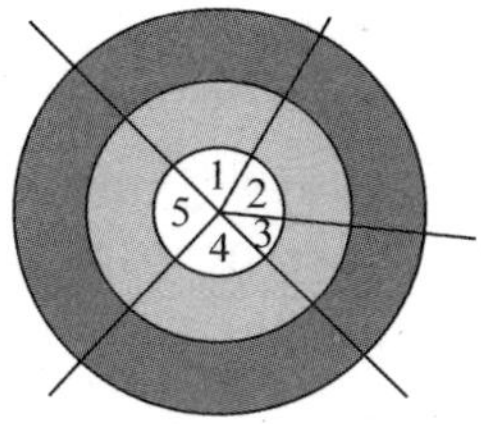

Figure 9.1 Structural analysis

In Figure 9.1, the first circle represents the arrangement of elements and its proportion; the second circle shows the cultivation of elements, including the lists of different influences that have an impact on elements and the relation between elements and the outside world etc.; the third circle lists application program and methods.

(2) Structure's effects of division and reunion. In real life, divisions and reunions of structures are widespread. They will attain good effects if handed well. Here I merely list some examples to bring out the theme.

1) Divide the whole in terms of structural unit. For instance: At a gathering, a patient says to a good doctor: "I extend thanks to you on behalf of all the patients." Actually, this sentence can be said by different patients: "I extend thanks to you on behalf of those who were ill, those who are ill and those who may fall ill in the future." Such an expression is much more vivid. Furthermore, we could use methods of replacements and exclusion etc. We may say, "It is unfortunate for those patients who didn't receive your treatment, while it is quite lucky for those patients to receive your treatment right now. I just hope that there will be more doctors like you who can bring benefits to the patients."

Here are some other instances: A large troop formation can be separated into a large number of guerrilla forces. A big task can be contracted into different parts.

2) Assemble the scattered things according to the rules of structure. For example, different corpsis are united to fight a war of annihilation; the equipment in different departments are gathered for a unified use, etc.

Structure determines everything and changes everything. The subject and the object have to be on the same level in order to adjust structure. The subject should have the capability and conditions to adjust the structure. If not, it can do nothing about it.

(3) Culture of joints. Being structure's connections, changes and synthetic effects, joints serve as thing's substitution, continuation and the focus of opportunity and challenge's coexistence. Things are universally contacted, which explains why joints are universal as well. Thus, everywhere is filled with opportunities and challenges.

Joints can be classified into the following kinds: periods of time, zone, inter-individual, the three dimensional lines, birth, vanish, transformation, the discontinuation of continuation, the fluctuation of market information, collocation of elements, unions of new couples, the making of decisions, so on and so forth. In a broad sense, every time and place is a joint in everything's development.

Joints pose opportunities as well as challenges. For example, "Taking Oath in the Peach Garden" and "The Scuba Diver" *in the Romance of the Three Kingdoms*, our daily talks, meetings and festivals in real life are typical joints filled with opportunities and challenges.

Let us take some phenomena for example. Borders are booming areas of bilateral trade, but they can also serve as places for two countries' conflicts. Tree-grafting gives birth to new species. The clashes between warm current and cold current generate thunder and lightening. The forest may lead to the replacement between oxygen and carbon dioxide; the power generation can be obtained through cutting by running cars on the streets and highways; the airstream in patio of high buildings can rise; the air-conditioning uniform can generate electricity through human activities; electricity is generated by the streams of tap and toilet, and so on. The above examples all serve as joints for further researches.

Chapter 10 Immortal Effects

Effect is the ending. It can not be deserted and it is imperishable. If you want to avoid a bad result, you must keep the effect theory in mind.

10.1 The assumption of the Immortal Effects

There is a famous law in physics, namely, "Substance is imperishable." It means substance will not perish but will just transform from one form to another. That's the immortality of substance. Wood is burned to ashes. Here substance is transformed from wood to ashes and smoke that is released to the air. Water is heated to be steam, which can cool off and turn into water again. Solid is heated to be fluid, which cool off and turn back to solid. This repeated cyclic change is pertinent on the immortality of substance. Just like substance, things' effects are immortal too.

10.1.1 The definition and classification of effects

(1) Definition of effect

Effect refers to results caused by substance's physical and chemical processes as well as responses and results brought about by all things and phenomena. As a result of constant development, effect is a physical form.

(2) Classification of effect

Effect can be classified into natural effect, plant effect and animal effect.

Natural effect refers to effect generated by the nature and substance, such as effects caused by galaxy revolution, thunder and lightning, earthquake, volcano, wind and rain, sea tide and river, etc. Other examples include heat effect, resonance effect, photoelectric effect, chemical effect and physical effect, and so on.

Plant effect refers to effect yielded by plant itself and its effects to the outside world, the birth, the change and death of plant, plant's ingestion and repayment to the nature, etc.

Animal effect refers to effect brought about by animals and their effects to the outside world, such as the birth, change, death, ingestion and repayment to the outside world of the animals, and so on. Animal effect includes human society effect and other animal effect. Human society effect is comprised of cultural effect, behavioral effect, thinking effect, star effect, celebrity effect and interpersonal effect, and so on. The impacts exerted by good or ill numbers, augury, belief, culture, traditions and customs all belong to cultural effects.

Effects can be subdivided into many small categories. For instance:

Key effect. In the process of accumulation, a little effect's role does not seem to be so obvious. However, when it comes to sudden change, a little effect often plays the essential role of igniter fuse, causing tremendous even essential change. That is key effect. On December 1, 1955, a black woman named Rosa Parks boarded a city bus after work. When asked to give her seat to a white by the bus driver, she, feeling exhausted, refused his request. The impact of her refusal has been expanded instantly due to the effect of other people. Black people in Montgomery city began to boycott buses in unity, which lasted for 381 days. Later, white people also started to go against the racial segregation system. In 1956, the supreme court of the USA declared the racial segregation system having violated the constitution. A petty black woman's refusal to give her seat should not have been something that mattered; however, it eventually overthrew an evil and entrenched social system.

Effects created by receiving others' favors. The reason why this effect shines comes from influences of the outside world's needs, favors or casual behaviors. Effects of benefits from others refer to those who neither made efforts on his own nor intended to achieve something but finally gain success. For instance, sitting in a car running on the highway, people would see beautiful scenes exhibited in front of them by trees alongside the roads; though one has not achieved any results, he gets quick promotion because of his personal conditions coinciding with the outsides' needs.

10.1.2 The reasoning of the Immortal Effect

(1) Any phenomenon's effect is imperishable. For example, photoelectric effect creates energy, influences other substances and consumes the nature.

Simultaneously, it releases discharges to the natural environment. This influence is everlasting and imperishable to the immortal universe, for the effect itself is a sort of substance too. Take some other examples: driving cars consumes energy and causes frictions to roads as well as produces automobile smoke. Body exercises would lead to expenditure of energy, cause frictions to the ground and the air, inhale oxygen and exhale carbon dioxide. Sitting quietly would exert pressure on the bearings, obstruct the air and disseminate scent. Spitting hits against the air, leaves discharges outside, and affects the environment. Human beings live to consume and to excrete. Public security bureau solves criminal cases by the scents; a man always leaves scents wherever he went by, which serves as an example of the immortality of effects.

(2) Analysis of people's behavioral effects. The effects of people's every behavior are immortal. For example, if you had done a good deed, first, it would have made an impact on the recipient, and he might have felt grateful to you; second, you would have contributed to the improvement of general atmosphere of society; third, you may somehow have felt losing something, though you might have received the baptism of love in the bottom of your heart and been motivated to be more helpful to others; fourth, the good deed itself exerts certain influence on the objective world as well as the cosmic space. If you had done something wrong, you would have made harmful or unhealthy impacts on victims and society, and added something unwholesome to the universe. Victims hate and you may feel quilty you bitterly, and you may feel guilty. It is also harmful to your psychological development direction and gives birth to some habitual development factors. It lays the possibility of leading you to self-destruction.

Supposed all the things in the universe were still, the universe would begin to move influenced just by the effect of a person's wink. In the later development and evolution, the effect of the wink is still there and will never vanish, in spite of the fact that the single effect doesn't have much influence and its immortal property doesn't appear so evident in the boundless universe where all sorts of effects are mixed with each other.

10.2 The properties of the Immortal Effects

10.2.1 Effect exists for good in the following effects influenced by it

As an effect comes into being, it creates and influences another effect and merges itself into another one. Someone may say that the effects of practicing singing and building up body are so good at one time, but why are the voice not so clear and the body so stiffening after a long pause? That doesn't amount to the perishing of effects, but the body's integration, offset and concealment of the original effect in the evolving physical and chemical effects, makes it seem not so obvious. The original effect begins to function and is merged into the body, making it impossible to eradicate them.

10.2.2 Effects set off effects

An effect could set off another effect, or even a large number of effects, which respectively lead to even more effects. A tiny effect's influence is tremendous as it develops constantly. Therefore, the occurrences of effects are interlinked and their influences are magnified and comprehensive. Thus, be cautious about the creation of unwholesome effects. Once created, it is rather difficult to clear up their influences. This can be demonstrated by the following popular sayings: "It is no use crying over spilt milk"; "An arrow will not turn back once it is shot out from the bow"; "It is easy to ask the God come but it is difficult to send the God away". At the same time, pushing for the creation of good effects actively, even tiny ones, will often change those seemingly impossible things. To a person, the efforts of moving towards right direction will always pay dividends.

10.2.3 Small effects could bring about big effects

A person should not feel that his power is nothing to mention. Actually, his power is immense, which is exhibited especially at the critical point of things' development. In 1963, American meteorologist Lorentz set up a mathematical model of atmospheric convection, employing computers to conduct numerical simulation on weather change. It is clearly seen that one thousandth error results in completely different weather pattern. A tiny little change would lead to enormously different results. Thus, Lorentz said exaggeratedly that a butterfly's several flaps of its wings in Brazil's tropical forests would lead to a tremendous tornado in Texas several weeks later. Scientists named this phenomenon "butterfly effect". Steam engine invented by Watt caused a worldwide industrial revolution; The reduction of memory chip's manufacturing costs triggers off the large-scale popularity of personal entertainment digital products—from mp3 to digital camera's popularity; Pirate software and pirate literary and artistic works become even more convenient with the appearance of disc copy devices and medium—the less costly reproduction and distribution technology make everyone potential plunderers of social civilization treasure; The appearance of condom caused sexual liberation in the USA; There is even a case in *A Dream of Red Mansions*①: "Sometimes by chance, a look or a glance may be one's fortune." According to the entrance examination system, mark determines everything, and a disparity of merely one point would lead to different courses of life; Mental disease formed in childhood would be severely magnified in adulthood; A small detonator could set off numerous explosives; A tiny breakdown of the spaceship could cause the failure of a giant project; A man without money is no man at all; A stitch in time saves nine; An ant hole may cause the collapse of a thousand-*li* dyke; A minimal error results in wild divergence; In the face of temptation, a momentary slip would put someone in prison or even makes him lose life. Just as a western ballad goes: For the want of a nail the shoe was lost, for the want of a shoe the horse was lost, for the want of a horse the rider was lost, for the want of a rider the battle was lost, for the want of a battle the empire was lost. Therefore, the devil is in the detail.

10.2.4 Cumulative effects

The influences of cumulative effects are immense. In *Romance of Three Kingdoms*, one of the most famous novels in China, Zhuge Liang is playing musical instrument in an empty and defense-free city. Suddenly the string was broken. This little act scares away Sima Yi's thousands of troops. This results from the cumulative effects of Zhuge Liang's usual maneuvers, not from the effect of a broken string. Likewise, people's habits and characters are results of cumulative effects. It is said that during a selection of recruiting people, Lin-

---

① It is also called *The Story of the Stone*. It was written by Cao Xueqin and it is one of China's Four Great Classical Novels.

coln mere took a rough look at an applicant and immediately turned him away. Someone asked, "Why did you turn him away by a mere look?" "I don't like the man's appearance", replied Lincoln. "But you can't judge a man by his appearance", someone continued. "Of course, a man can't choose his own appearance. But he must be responsible for his countenance after becoming 20", Lincoln replied. This example points to cumulative effects: Being accustomed to peacefulness and kindness, someone is bound to be calm and happy. As time goes by, his face is always brimming with kindness. A person, harboring evil intentions and scheming against others, always appears in a poker face. As time goes by, it develops into untruthful face. Being used to hostility and evils for a long time, one tends to have the evil countenance stereotyped on his face.

10.2.5 The causality of effects

An effect can be traced back to its original effect. "Everything in existence is reasonable." Everything happens for a reason.

Effect is always intimately related with cause and there is no exception. Certain destruction and change to the natural world would bring about retributive effects, such as over-exploitation of subsurface resource causes the crisis of our living environment. Pollution and destruction of the atmosphere has negative impacts on the climate, etc. To a man, the retribution refers to reactively negative impacts produced by his own effects or his relatives. If a man has done something wrong, just as mentioned above, first, it exerts a range of impacts. Its retributions include: first, it initiates and has cumulative effects on the development trend of bad habitual behaviors; second, it is the microwave effects of victims' retributive threats and curses; third, being in a state of anxiety, one would have an uneasy conscience that keeps disturbing him; fourth, brain microwave resulting from the act including all sorts of disturbing microwaves before, amid and after wrongdoing are released into the outside world. His children, grandchildren, close relatives and offspring are susceptible to those microwaves that may bring about immensely unhealthy effects. He must be punished by others if not by heaven, by his conscience if not by others. He who is indebted to others must pay back what he owed either by himself or by his offsprings. "Good begets good and evil begets evil. God's mill grinds slow but sure." Change is followed by long-term thoughts, what one has in mind will rebound upon himself. One should bear in mind that what he says and does will visit on him. Therefore, good can be distinguished from evil.

Effect follows each other in order, and cause and effect matches with each other. A person's childhood experiences will affect his adulthood or even his whole life. Prior experiences have influences on subsequent behaviors. The pioneer's experiences affect the followers' behaviors. Cause brings effect, and every effect results from cause. Sigmund Freud, the founder of psychoanalysis believes that, children's growing environment and experiences in early years play essential roles in shaping their character in their adult years. Many adults' abnormal psychology and psychological conflicts can be traced back to their early-year traumatic experiences and repressed complex. He also thinks that human psychological activities have strict cause-effect relationship. Nothing happens accidentally. Dream is no exception. Instead of being accidental associations, dream serves as the satisfaction of desires. During sleeping, one's self-inspection becomes loose; desires in subconsciousness break into consciousness in disguise, forming dream. Thus, it can be seen that dream is a euphemistic expression of desires in the subconsciousness that have been restrained while being awake. Dream serves as the secret path towards subconsciousness. By analyzing dreams, we can glimpse into one's psychology and probe into desires and conflicts in subconsciousness. Neurosis can be cured through the explanation of dreams.

One is incurable once implicated by what his relatives have done. Retribution would incriminate those who are nearest and dearest to the evil-doer. As the proverb goes, "One who stays near vermilion gets stained red, and one who stays near ink gets stained black." Similarly, those who are close to good people are fortunate while those who are close to evils will suffer. As a sort of corporal punishment in Chinese history, the incrimination of nine generations of a family demonstrates how unfortunate those who are near evils can be, although the evils here are not necessarily true evils. Yu Ji① in "Farewell to My Concubine"② is a kind person, but she only put an end to her own life

---

① Yu Ji is the concubine of warlord Xiang Yu. She committed suicide with Xiang Yu's sword.

② Xiang Yu fought with Liu Bang, and Xiang Yu was surrounded by Liu Bang's forces and was on the verge of total defeat. Xiang Yu called his favorite concubine to run away, but Yu Ji beged to die along her master.

for being implicated by Xiang Yu[①]. So was Yang Yuhuan[②], who was hung at Mawei slope for the misdeeds of her elder brother Yang Guozhong and her husband Li Longji.

Cause and effect relationships can be divided into the absolute one and the relative one. From the perspective of all effects in the whole universe, cause and effect relationship is absolute, allowing for no contingency; from the standpoint of partial effects, cause and effect relationship is relative. The cause and effect relationship we talked above is relative. Good people are not necessarily to get good returns. People know that from their experiences. For instance, are victims of disasters like air crash, earthquake and tsunami all bad people? Are good people immune to disasters? We know those are not true. Dying, suffering miseries and setbacks as well as windfall gains result from the combined effects of themselves as well as the outside world. Viewing from the whole universe, those happenings are inevitable and predestined.

10.2.6 Positiveness and negativeness of effects

Effects are separated into positive and negative effects. Their disparity lies in the different directions of effects' action. At one point of things, meetings of positive and negative effects would result in the counteraction and submergence of effects. It doesn't amount to the disappearance of effects. Instead, it means the direction of effects' influences has been changed. New effects are created and the original forms of effects have been changed. Dialectical materialism thinks that everything has its advantages and disadvantages. Non-living existence yields natural effects, namely effects created by nature. Living existence produces natural effects and active effects, both of which have positive and negative effects. Effect itself has neither positiveness nor negativeness. In the analysis of two or several opposite effects, people tend to add their own emotions and divide effects into positive effects and negative ones. To those evildoers, their good deeds will neutralize the negative effects that have been created to their bodies. Nevertheless, the original effects will never vanish.

---

① Xiang Yu is a prominent warlord in the late Qin dynasty. After the fall of the Qin dynasty, Xiang Yu proclaimed himself "Overlord of Western Chu".

② Yang Yuhuan is known as one of the "Four Beauties" of ancient China. She was the beloved consort of Emperor Xuanzong (Li Longji) of the Tang Dynasty during his later years.

10.2.7 The appropriateness of effects

Excessive amount of effects would overflow; lack of effects would be useless. That is, the specific excessive amount of effects far surpasses the object's capacity; thereby it does damage to the object or influences other objects, bringing about other problems, normally negative impacts. However, the deficient amount of effects would not achieve the desired result so as to make every effort futile.

To the natural world, over-exploitation of natural resources is like a gravestone that human being has excavated for themselves, while under-exploitation can not meet human's needs. To a man, the overuse of his brain which exceeds what it can stand would damage his brain; the effect would spread to other parts of the body and brings in illness. If one doesn't utilize his brain to the fullest, it will lead to the other consequence. Brain cells and intelligence would wither; one will be more prone to senile dementia. Likewise, excessive eating and exercising too much have the same effects. One should allow others to express different opinions and be selective in hearing and taking advice. He who merely gives in to different views will get nowhere; not paying much attention to advice is far from impartial. It is even more inadvisable to merely suppress different ideas, for others will not be sincerely convinced by suppression. Just as the famous Chinese saying goes, "The water can not only bear the boat but also overturn it." In nurturing children, both spoiling and lack of love are damaging.

In sum, substance, its cause and effect, inborn law as well as effects produced by substance are all imperishable. Therefore, it can be concluded that everything in the universe is imperishable.

10.3 Sources of effects

The everlasting time is set against transience. While being meaningful to shortness, time is meaningless to permanence. Thus, time becomes meaningless to the whole universe. To human beings, sources of effects are classified into three types, namely, past, present and future. The influence of sources of effects is towards forward, namely the orientation of future. These three sources of effects exist simultaneously. While effects of sources of the past is basically transformed into the present's sources or they continue to spread to the future, effects of sources of the past is still alive than dead, giving off more or less effects. Though the sources of effects of the future have not exerted any influence yet, it already took shape. The sources of effects of the present are the most direct, active and vital sources, which directly affect the present and the future. Being the present's blueprint and plan, the sources of effects of the future are gathering of heaven, earth and human beings' efforts. As a naturally formed pattern, the future's effects exist objectively and for real. When we are still in the present moment, the sources of effects of the future have already been formed. Like the belt of vehicle, the sources of effects of the future have been fixed in advance. Information wave of mind including planning and contemplating, once given off, would conduct comprehensive analysis and accommodation and thus form consequences naturally. "Man proposes and good disposes." "Chance contrives better than us." What's heaven? Heaven is the laws of development of things and the comprehensive forces of subjective effects as well as objective effects. No matter how accurate man calculates, it can not fully conform to the law. Sometimes it even violates the law and can not be freed from the bondage of combined effects of the whole environment.

The past generates the present, which then yields future. Effects influence things occurring later. Actually, they have impacts on the future. The information waves given off by self-esteemed and strong-willed people are strong and powerful; therefore, they have many followers and are more adaptable to the environment of the future and get more chances of success. Since the effects created by the three sources of effects are oriented towards the future and the increases of effects as well as their influences are at accelerating pace, society is developing faster and faster.

10.4 Predicting the future

(1) Effects are predictable. As long as the above sources of effects are physical and predictable, the future is predictable. Being the most acute and advanced substance, human's brain has the trait of predictability. It has premonition in the daytime and predictable dreams at night.

(2) Predicting the future. One's future results from the comprehensive action of his effects and external effects, so are the development prospects of things and celestial bodies. Therefore, whether a man succeeds or not isn't wholly dependent on himself. It is not necessary for good people to have good

returns, but bad people are predestined to get no good return. Disaster is the encounter between man and the negative influence of external effects, while luck serves as the meeting between man and the positive influence of external effects. For human beings who have no control over the universe, the situation there is volatile. Any successful experience and self-satisfied idea may incur losses.

10.5 Law of universal response

Everything in the universe generates effects, which will definitely create responses to the outside world. Everything has and receives responses. Sources of responses are separated into static response and dynamic response. Response, in essence, is effect.

People have a great variety of responses. a. Responses of feelings. Others' feelings would be conveyed to you. Likewise, your feelings would be conveyed to others as well. The receivers tend to have receptive responses, repellent responses and indifferent responses. They will accept you when sympathizing with you, repel you when feeling hostile towards you and be indifferent to you when feeling impartial to you. b. Responses of words and deeds. Others' words and deeds will generate responses within you. Perhaps, you will be in sympathy with them, at ease with them or against them. c. Responses of attitudes. The signals of attitudes you gave to others would affect or change others, vice versa. d. Other things in the outside world would exert influences (responses) on you.

Nature has its own laws. Resorting to the interplay of all things, nature seems to be changing unrestrainedly and unrestrictedly. "All things are free to grow without limits." Animals have their circles and people have their own circles. Members within a certain circle are closely bound up with each other, interacting with each other, influencing each other and responding to each other. Therefore, any member within the circle at least has the obligation to safeguard the circle's common interest. If he fails to figure out that the basic reason or to fulfill the obligation, he doesn't deserve to be a member of the circle.

10.6 Spirit is physical

Spirit is physical, serving as a special matter generated by substance.

First, the information source of spirit comes from the objective physical world; second, the physical world stimulates the brain that then yields spiritual substance following the inheritance; third, thought's process and result is the process of new substance's creation and release. The form of spirit that we usually know is only an external form of the spiritual substance. Spirit exhibits itself to the outside of brain in the form of wave. The spiritual wave to the universe is like a shock wave to the universe or a drop of water to the ocean. Its effect will never vanish. So the effect of an idea or a dream is imperishable, too. Spirit is imperishable. With the development of science, we will invent the equipment that can measure spiritual waves and be perfectly clear about thought's form and content. Sooner or later, people won't have to study anymore. Just by putting spiritual substance or chip-like substance, people will acquire relative knowledge base.

Man's behaviors, including affectional behavior, mental activity, etc. , all have sources. They are all reflections of substance. The root lies in gene, a substance too. Concept is subject to substance. Substance, including brain substance, yields concepts in the course of its development. Without substance, there will be no concepts at all. As a sort of substance, thought is a special substance created by the brain.

Different kinds of animals have distinctively different intellectual conditions. The reason lies in the different intellectual substance in brain. Let's take elephant and cheetah for example. Their intelligence and agility are never the same. The intellectual conditions among people are not only vastly different but also have distinctive hereditary features, which prove that their spiritual substances are different.

To sum up, spirit is physical. Apart from the solidified spiritual substances such as books, the foundation of brain as well as brain's thinking process and result are all physical. The radiation wave to the external world from brain's reflections is substance and its effects are imperishable. If there were gods in the universe, they must be substance too, which is an advanced substance. According to our normal thinking, the universe has nothing else besides substance. In reality, the universe is made up by substance and non-substance. The opposite side of substance is non-substance instead of spirit. Everything in the universe develops in accordance with certain laws. Therefore, any result is predestined.

10.7 Necessitarianism of effect

The universe has only certainty rather than contingency. All the things are predestined: the waxes and wanes of the moon and stars, growth of grass and singing of insects on the earth, people's blinking and pronouncing of words, colorful pictures and brilliant sound of television, midges' flying, mosquito bites and hunting dogs' chasing after hares. All the tiny changes in the world are certain. Supposed we were able to observe all the effects in the universe as clearly as we observe a bowl of clear water, the certainty would become obvious; if we record every change in the world, we will find out the source of every change later and let all things in the universe to continue to change in their own way. They will still abide by the law of self-evolution and combine with the influences of the external world to develop according to the destined. We can summarize it with four sentences: All things in the universe yield effects. All things in the universe develop under the push of effects, for all the change is effect itself and its result. The generation of effects complies with cause-effect law; a certain effect will lead to a certain result (namely effect or influence); all effects are conducted according to certain law and way and everything is destined. All the destined lead to all the certainty. To summarize again: all the things in the universe including spirit are substances and all the substances develop in accordance with certain law (namely destined). Therefore, any subsequent results from them (namely effect) are predestined. The reason why we believe there are contingencies in the happenings of things is because we don't stand tall enough to see the whole picture and we fail to observe and recognize the development of all effects. When unexpected phenomena occur, they take us by storm and thus generate the concepts of contingency and certainty within our hearts.

Do not judge a person by his success or failure. Since the result of a thing, the ending of one's life and one's success or failure are all dependent on combined actions of its own effects and external effects. One doesn't have the total guarantee, control and choice over these things. Thus, status, fortune, fame, etc., are not the criterion to judge a person. The only right criterion to judge a person is to see how he does, that is how to behave himself. Therefore, when we are asked to judge on a matter, we should see the result; when asked to judge over a person, we shall see the process he presides over things. One's influence has two directions: a. to the external world; b. to himself. Pursuing happiness is one's lifelong goal. Benefiting oneself at others' cost would lead to the loss of one's moral quality. Thus, the ideal life should be like this: One feels no regrets over his treatment of others and he treats himself without suffering losses.

One's destiny depends on oneself. "Certainty of effect" is objective and indisputable. Thus, since given facts are certain to happen, one has to accept them calmly and try to comply with them. However, he must make strenuous efforts to fight for his future. That is because on the one hand, objective effects have become destined and the impetus of fighting has been mobilized by his accumulative effects. On the other hand, under the condition of permanent objective effects, as long as one's subjective effects get larger, the sum total of one's subjective and objective effects, inductive to one's development, becomes larger accordingly; one's chances of success and good fate are increasing. Therefore, the hope and flux of one's future lies in oneself. Only by adding new positive effects can ideal effects be further developed. Success is the inevitable result of accumulative effects. What happened in the past and will occur in the future are certainties of effects. Nevertheless, what hasn't happened yet is changeable. Thus, it is said that what occurred is certain to have happened, but things of the future can be changed. Being struggling is sure to create positive effects and being demoralized is certain to generate negative effects. One should not only utilize initiative but give full play to his initiative. He should also take control of his own destiny, striving for his destiny, since effects are imperishable and their actions exist all the time. Hard work might not ensure success, for it is the result of synthetic action of subjective and objective effects. However, not working hard means no chance of success at all. Act aggressively, do not expect too much. To improve chances of fulfilling his own effects, one has to try his best to expand controllable scale of the external world and strengthen his control over his own effects.

Necessitarianism and fatalism have essential differences. Fatalism believes that things' change and development as well as man's life or death and poverty or wealth is all unpredictable and unforeseeable. They are decided upon by unknown power and therefore are unchangeable. That power is filled with the content of mysticism and God. The main differences between necessitarianism and fatalism are: first, fatalism belongs to idealism, while necessitarianism belongs to materialism; second, all things in the universe develop by triggering off effects. Certain effect will lead to certain result without being determined by God; third, the synthetic action of subjective and objective effects we talked above is knowable and predictable, without any mystery. Nevertheless, today's scientific and technological level hasn't reached up to all-round comprehension yet.

10.8 The follow-up of moral education

Due to some distorting phenomena in the society, some people began to have unbalanced thoughts in their mind. It further develops into mental diseases of being afraid of others' achievement, wishing others live worse lives than he does and looking for chances to stir up trouble. That's why lots of people merely look at the endangered victims and would not offer a hand. Even some onlookers gather together to create a disturbance and laugh at those who rise gallantly to the occasion. These phenomena demonstrate the morbidity of those onlookers. We should guide people to make money through proper channels by means of regulations, public opinion and other measures so as to balance their mind and advocate high morality.

We could not ignore self-interest, the mentality with which we are born. Morality can not be built up as long as benefits (including spirit and material) of abiding by morality are much lower than benefits of ignoring morality or abiding by morality could even spark off disasters. Thus, moral education should adapt to self-interest of man's nature, instead of going to the extreme to build castle in the air. These acts demonstrate "imperishable effect". Imperishable effect should be taught in modern moral education. It is for sure that one must absorb the effects he created.

There is a story which deeply reveals the truth of imperishable effect.

> At the age of 28, Phyllis, a black man, became the champion of European long-distance race. To commend his contributions to the country's sports, his motherland showered him with great honor.
>
> Once, he was invited to deliver a speech in the country's largest prison. Facing thousands of various criminals, Phyllis talked about his underprivileged childhood and how he struggled unremittingly to change his destiny in difficulties. The speech gave rise to a lot of responses among the audience and won prolonged applause.
>
> As he finished the speech and was prepared to step off the platform, a criminal rose up and asked him a question, "Whom do you thank the most in your life?"
>
> Instead of giving the answer immediately, Phyllis sank into a deep thought. At the

same time, the questioner popped out several questions in succession. "Is him your coach or your favorite parents or your best friend..."

Phyllis kept shaking his head and said, "I feel much grateful to all of them, but they are not the person to whom I feel the most grateful. It is a special person to whom I feel the most thankful to. For, without him I would not have been who I am now. I really hope that I could find him and give him thanks personally."

Who is him? Everyone was waiting for the answer.

"It was the stealer, who stole my bicycle many years ago, to whom I thank most," replied Phyllis. That made an outcry in the audience: The audience was wondering how it was possible that Phyllis feels the most grateful to a stealer.

Then, Phyllis told the whole story to the surprised audience: At the age of 13, he was a grade one student in middle junior school. His family was so poor that he couldn't afford to live in school. Therefore, he had to ride a run-down bicycle to shuttle between his house and school, which were 5 kilometers long. It was an extremely old bicycle. Nevertheless, one day, he forgot to lock his bicycle after he arrived at school and the bicycle was stolen.

His family couldn't afford to buy him a new bicycle, so he had to run to school. He ran 10 kilometers every day and bore lots of hardships in the following three years. However, the daily running at first made him the champion of school long-distance race and then helped him get a second in his city... Thereafter, he became a career sportsman. He said honestly, "I would not have achieved what I have today if not for the loss of my bicycle." He always felt sorry for the stealer, "Perhaps it was his first stealing. Owing to carelessness, he got the bicycle so easily and became all the more out of control from then on."

The audience burst into laughter and said that such a coincidence would not occur. However, just at that moment someone in the audience began crying and cried out, "I am the staler!" The staff of the prison roared at him, "Hikulay, what the hell are you doing? Do you wanna add more to your prison term?"

Phyllis stopped the staff and listened to the criminal's nagging patiently. As Phyllis heard the criminal saying how he assembled three types of parts of the bicycle and its crossbar had welding marks as well as its back tire had broken marks, Phyllis stepped off the platform silently. He picked off a medal from his chest and pinned it on Hikulay's chest. Feeling shamed, Hikulay kept repeating the words, "I am not worthy of it. I am not worthy of it."

Phyllis left and said he would come back.

People felt amazed, "It must be God who fixed it." Loser got the most munificent present while receiver lost the most treasured thing. Gain and loss are just like a couple of brothers who depart on forks in the road. The longer distance they cover, the more different they become.